# PHYSICAL PRINCIPLES IN SENSING AND SIGNALING

D1758939

# Physical Principles in Sensing and Signaling

**With an Introduction to Modeling in Biology**

ROBERT G. ENDRES

*Division of Molecular Biosciences, Imperial College London*

OXFORD
UNIVERSITY PRESS

**OXFORD**
UNIVERSITY PRESS

Great Clarendon Street, Oxford, OX2 6DP,
United Kingdom

Oxford University Press is a department of the University of Oxford.
It furthers the University's objective of excellence in research, scholarship,
and education by publishing worldwide. Oxford is a registered trade mark of
Oxford University Press in the UK and in certain other countries

First Edition published in 2013

Impression: 1

British Library Cataloguing in Publication Data

Data available

Library of Congress Cataloging in Publication Data

Data available

ISBN  978–0–19–960063–2 (Hbk.)
        978–0–19–960064–9 (Pbk.)

Printed and bound by
CPI Group (UK) Ltd, Croydon, CR0 4YY

*To Sofia and Elena*

# Preface

This textbook is aimed at newcomers in the field of biological physics, especially motivated undergraduate and beginning graduate students with a mathematical and physical background, taking interest in the area of quantitative biology. The unique feature of this book is its conceptual, informal, and friendly introduction to the topic, aiming to translate new developments in multidisciplinary science to a wide audience. At the expense of some rigor, the book mainly teaches concepts and intuition by directly applying quantitative methods to biological problems. It focuses on simplified cases to avoid complicated mathematics as much as possible.

Topics covered include the accurate sensing of chemicals by cells, signal and information processing, similarities between biology and engineering, and the issue of noise in the reliable working of cells. For additional guidance, all of these topics are connected, and mostly explained in terms of the chemotaxis signaling pathway in the bacterium *Escherichia coli*. This pathway includes cell-surface receptors and a signal transduction mechanism, which regulates flagellated rotary motors, ultimately allowing cells to sense and swim towards nutrients and away from toxins. The pathway is regarded as a paradigm of sensory pathways, and is considered an excellent starting point for further studies.

The origins of this book lie in a final year undergraduate module in Integrative Systems Biology, as well as lectures from the Masters in Research course in Systems and Synthetic Biology. I taught in both courses to a colorful mixture of students at Imperial College London. The challenge was to be able to communicate advanced material, in part quite theoretical in nature, to students who often lacked the proper technical background.

Communicating science to a wide audience, including potentially non-scientists, is an important but often difficult task, requiring generally underdeveloped skills. Science is becoming increasingly multidisciplinary, merging the disciplines of biology, physics, chemistry, engineering, mathematics, and computer sciences. This is reflected by the research funding agencies, which place more and more emphasis on science to affect the lives of everyday people, and to reach the general public more effectively. With this book, I hope I will be able to play a part in this challenging endeavor.

This book would not have been possible without the help and support of many others. Specifically, I very much thank Professor Rob Phillips (Caltech, USA) for his friendly and humorous support during the planning and writing of the book, and for his critical feedback. I also thank my former student PhD Diana Clauznitzer, who did a lot of the underlying research and provided figures for the book. Furthermore, I thank my wife, Erin Ingrassia, for her loving support and proof-reading of the manuscript. Importantly, I would like to acknowledge Professor Ned Wingreen (Princeton, USA) who introduced me to this exciting line of research and taught me much wonderful biological physics, and Professor Victor Sourjik (University of Heidelberg, Germany) for providing me with excellent data and a fruitful collaboration over the last few years. Special thanks go to Professor Howard Berg (Harvard University, USA) for generously providing the experimental cover-page image of swimming bacteria. A lot of the described phenomena are based, in one way or another, on his pioneering work, stretching impressively over several decades. Finally, let me acknowledge all the exciting research publications, which shaped this book. The most relevant publications are listed at the end of each chapter but many others, especially on the experimental side, are omitted. I hope the reader can excuse this shortcoming, which made this book short and manageable to read and write.

London, April 2012

# Contents

# Introduction

1

Although invisible to the naked eye, bacterial cells are large enough to make complex decisions. Cells are composed of thousands of different molecular species including DNA, proteins, and smaller molecules, allowing them to sense their environment, process this information, and respond accordingly. Such responses can include chemical reactions, e.g. expression of genes, or some kind of force generation, e.g. to control of cell movement. Despite these properties, a living cell exists in the physical world and follows its laws. Keeping this in mind can help answer questions such as how cells work and why they implement solutions to problems the way they do. Specifically, applying physical principles in biology allows researchers to solve challenging problems, including how accurately biological cells can sense chemicals in their environment, how they amplify signals and adapt to persistent stimulation, or how they encode physical stimuli in biochemical pathways. In this book, the reader is introduced to this fascinating topic without extensive prior knowledge in biological physics.

In simplest terms, a living cell is equipped with cell-membrane receptors, which allow the cells to sense a wide spectrum of chemical molecules in their environment, such as nutrients, toxins, and molecules secreted by other cells. These molecules (ligands) bind to specific receptors, which generally triggers a conformational change in the receptor proteins, hence transmitting the stimulus through the membrane into the cell interior. Similar conformational changes can also be induced by changes in temperature or pH. Such changes can subsequently be read out by cell-interior (cytoplasmic) proteins, which may prefer binding to the new receptor conformation. The "message" is then passed on to other proteins inside the cell, defining a biological pathway. Besides sensing, chemicals can also enter the cell through pores and transporters in the cell membrane, and subsequently affect processes inside the cell. Ultimately, stimuli lead to changes in gene expression, metabolism, cell size and shape, or motility.

Biology is quite messy, as cells constantly exchange material and energy with their surroundings. These properties make understanding them very different from traditional physics problems. Physics has been extremely successful, since its problems can often be studied in isolation, and either involve very small (e.g. atom, molecule, star, planet) or very large numbers of objects (e.g. gas, fluid, solid, galaxy), allowing application of powerful physical theorems. There are also a number of exactly solvable problems in physics, and the remaining ones are often extrapolated from these. Maybe not surprisingly, biology and physics historically went different ways.

Unlike physics, biology was often rather descriptive, stressing the importance of molecular details and the variety of seemingly different observations. However, this has changed over the last two decades or so, as biology has become more quantitative due to new imaging, high-throughput analysis, and sequencing technologies. The vast amounts of quantitative data created a demand for the application of statistical and mathematical analysis tools. To understand the data, models are crafted and solved, their predictions compared with new data, and our understanding enhanced.

For such modeling, the processes in a cell are often conceptualized as networks of interacting molecules such as genes and proteins. Although useful tools, such networks obscure the fundamental link between physics and biology. Without a doubt, biological systems have evolved under the constraints of physical laws and are optimized to sense and encode physical stimuli, encountered by the cell with the statistics of a given environment. Physical stimuli are traditionally thought to be soluble chemicals of the fluid environment, which bind receptors and activate pathways. However, what effect the physical nature of the stimuli has on the process of sensing itself is often neglected. For instance, ligands arrive at the cell surface randomly by diffusion, making the process of measurement very uncertain at low ligand concentrations. Additionally, physical stimuli include forces and shear stresses exerted by the surroundings, neighboring cells, or fluid flow. These effects are major determinants of cell development, such as in the case of stem cells. Hence, all these examples point towards the need for a physical analysis of biological phenomena.

There is another caveat about modeling and understanding biology. Biological pathways consist of chemical reactions in the crowded environment of the cell interior. In fact, there are hundreds of different pathways working simultaneously at any given moment in time, and avoidance of unwanted cross-talk between these appears a difficult task for the cell (or at least for us to grasp). As a result, biological pathways are quite different from hard-wired electronic circuits. Nevertheless, living cells sense their environment, process this information, and react accordingly. Hence, a cell has to cope with the same engineering problems that humans dealt with at the onset the electronic era. Perhaps not surprisingly, we sometimes find that biological pathways are designed by evolution similarly to how an engineer would design solutions of analogous problems in our macroscopic world.

In this book we chose chemotaxis in the bacterium *Escherichia coli* as a case study to exemplify the importance of physical end engineering principles in understanding biological processes. Chemotaxis allows cells to sense and swim towards nutrients and away from toxins. Importantly, *E. coli* is highly sensitive to small changes of chemical concentration in their environment over a broad range of background concentration. This process is particularly well characterized in *E. coli* by experiments, and hence chemotaxis in this bacterium is suitable for further quantitative analysis. While *E. coli* has been chosen as a model organism for historical reasons, it is worth noting that there is really nothing special about this bacterium, nor is it general enough to be representative of others. Briefly, *E. coli* lives in our gut and makes, together with all our other bacterial inhabitants, approximately 5% of our body weight and 90% of our body's cells. This bacterium helps symbiotically digest our food, but can also live outside our bodies in soil and water. (There are also variants of *E. coli* which cause diarrhea and food poisoning.) In 1997 it was among the first species whose DNA was sequenced; that is, the order of the bases A, C, G, and T was determined. Subsequent analysis indicated that the DNA sequence encodes about 4,000 genes, stretches of DNA sequence which translate into proteins. These proteins can do certain things, e.g. make more DNA or protein, enzymatically digest food, or allow the cell to recognize chemicals in its surroundings.

This book is organized as follows. Chapter 2 introduces chemotaxis in *E. coli* at a phenomenological level, and describes how the pathway allows these cells to sense, signal, and swim towards chemical attractants. These observations are placed in the context of quantitative experiments conducted on living cells. Chapters 3 and 4 describe respective important physical concepts and mathematical tools, necessary for understanding the material to follow. Chapters 5 and 6 provide explanations for two of the most remarkable features of this pathway. Specifically, Chapter 5 covers signal amplification and integration by cooperative receptors, and Chapter 6 explains the origins of precise adaptation to a persistent stimulation by the cells. Chapter 7 illustrates how receptors localize in the cell, form clusters, and cooperate. Chapter 8 describes the most important cell external and internal noise sources, which negatively affect sensing accuracy and signaling fidelity. Chapter 9 delivers the engineering principles of the whole pathway down to the motor, specifically treating the pathway in terms of amplifiers and filters known from electrical engineering. Chapter 10 concerns optimality in the chemotaxis pathway, and hence discusses fine tuning for further improvement of the pathway performance. Chapter 11 presents sensing and signaling from the perspective of information transmission. In Chapter 12 we place *E. coli* chemotaxis in the broader context by illustrating similarities and differences to other bacterial species and pathways. Finally, in the Appendix we cover more techniques used in modeling biological phenomena and in analysing data.

# Chapter summary

- Although not conceptually different from non-living matter, living biological systems are open (exchange energy and mass) and hence are "messy" to deal with.

- New measurement technologies in biology demand mathematical analysis and modeling to make sense of the data.
- Physical principles can help when thinking about biological problem to simplify biological complexity and to constrain models.
- Cells sense not only chemical but also physical stimuli, including shear and forces.
- Cells solve sensing and signaling problems surprisingly similar to us in the macroscopic engineering world.
- Chemotaxis allows bacteria such as *E. coli* to sense and swim towards chemicals. The chemotaxis pathway is a well-characterized system, and was used early on to gain quantitative insights into a biological problem.

# Further reading

(1) *What is Life?*, by Erwin Schrődinger (Cambridge University Press, 1944).

(2) *Random Walks in Biology*, by Howard C. Berg (Princeton, 1993).

(3) *Physical Biology of the Cell*, by Rob Phillips, Jane Kondev, and Julie Theriot (Garland Science, 1st Edition, 2009).

(4) *Essential Cell Biology*, by Bruce Alberts, Dennis Bray, Karen Hopkin, Alexander Johnson, Julian Lewis, Martin Raff, Keith Roberts, and Peter Walter (Garland Science, 2009).

# Chemotaxis in bacterium *Escherichia coli*

**2**

Chemotaxis is the phenomenon that cells such as *Escherichia coli* bacteria swim up chemical gradients of attractants and down chemical gradients of repellents. The biochemical chemotaxis pathway equips these cells with remarkable sensing and signaling properties, which need to be unraveled if we want to understand how cells perceive their environment, compute, and make decisions. These remarkable properties include the ability of these cells to detect minute changes in chemicals with high sensitivity, and to adapt to persistent stimuli so that they exhibit only a transient response after which they behave no differently from before the stimulus was applied. Furthermore, chemotaxis works extremely well over a wide range of chemical concentrations (about five orders of magnitude). How do these cells sense so accurately, adapt, integrate different and potentially conflicting stimuli, and decipher noisy, fluctuating signals?

Despite our focus on *E. coli* chemotaxis, many of our results are rather general, and will apply to other types of chemotactic cells as well. In fact, all cells are able to sense their environment. If motile, they can migrate towards favorable and away from unfavorable environments. In addition to bacteria, examples of chemotactic cells include higher single-cell organisms. In the soil, amoeba sense chemicals secreted by bacteria and hunt them for food. Budding yeast sense mating pheromone and grows towards nearby mating partners for sexual reproduction. As part of our immune system, white blood cells chemotax towards locations of infections. To learn and memorize, the neurons in our brain form growth cones (subcellular protrusions) to make connections with other neurons. Hence, chemotaxis plays fundamental roles in many areas of biology, making it a key area of quantitative biology. However, immobile cells also benefit from knowing their chemical environment. An important example is embryonic development, during which cells read morphogen gradients. This allows them to locally express genes for segmentation of the embryo and the development of the body parts.

## 2.1 Chemical gradient sensing

There are two main kinds of gradient sensing. Fig. 2.1 illustrates how large surface-adhered cells can reliably sense spatial chemical

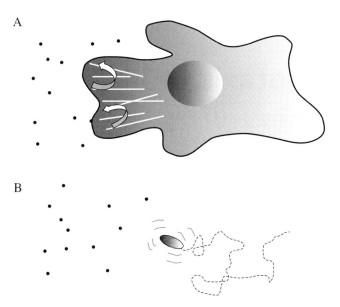

**Fig. 2.1** Two types of chemical gradient sensing. (A) Large, nucleated cells perform direct spatial gradient sensing, where gradient direction can be resolved across length of cell. Enhanced ligand-receptor binding leads to amplification by positive feedback (curved arrow) and enhanced polymerization of a cytoskeletal protein (linear lines) to push the cell forward. (B) Freely swimming bacteria suffer from rotational diffusion, which corrupts their directional measurement. The preferred method in this case is temporal gradient sensing.

concentration gradients across their cell length in a relatively short amount of time. In contrast, freely swimming bacteria are believed to be too small for a direct reliable measurement of the gradient. In particular, any directional measurement of the gradient suffers from rotational diffusion of the cell, randomizing the cell's orientation. Instead, small bacteria measure gradients in time while swimming. Using a random-walk type swimming strategy, they can bias their movement in a favorable direction. From all the bacteria, the chemotactic properties of *E. coli* are best characterized. And due to its relative simplicity, this bacterium's chemotaxis pathway is the ideal model system. In short, the chemotactic movement of *E. coli* works as follows: cells can either undergo "runs", straight swims of about 1 s duration, or "tumbles"—erratic movement of about 0.1 s for random reorientation of the cell body. If cells happen to swim in the right direction of the gradient, runs are long on average, and if they swim in the wrong direction of the gradient, runs are short on average, and cells quickly choose a new and hopefully more favorable direction of swimming.

## 2.2   "Nose and brain": the receptor cluster

To sense molecules in their environment, bacterial cells are equipped with different types of chemoreceptos, which can bind attractants (nutrients such as sugars and amino acids) and repellents (e.g. certain metal ions). In addition, they can also sense pH, oxygen, and temperature. Chemoreceptors exist in hundreds to thousands of copies, and cluster predominantly at the cell poles. As illustrated in Fig. 2.2, the polar receptor clusters are both the cell's "nose" and "brain" for regulating their behavior. How are the receptor clusters regulated and made by the cell?

A

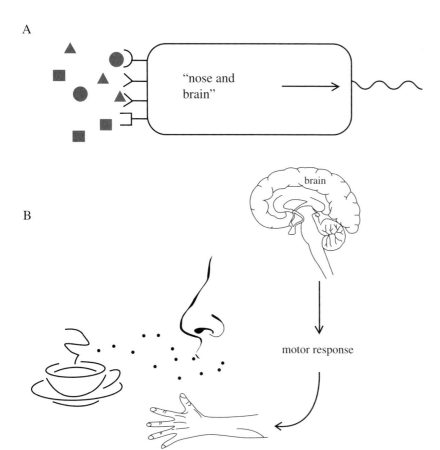

B

Fig. 2.2 From chemosensation to motor control in bacteria (A) and humans (B). While bacteria do not have a brain (obviously), they compute with their receptor cluster and its associated proteins.

Fig. 2.3 shows that receptor genes are transcribed from the double-stranded DNA to single-stranded messenger RNA (mRNA) by RNA polymerase. These mRNAs are subsequently translated into amino-acid sequences using ribosomes. Subsequently, the amino-acid sequences are generally folded into their corresponding 3D protein structures. Chemoreceptors, however, are simultaneously inserted in the inner membrane by the secretion machinery. Once folded in the inner membrane, they first form receptor dimers, then trimers of dimers, and finally tend to localize at the cell poles. Also shown are pores in the outer membrane and transporters in the inner membrane for amino acids and sugars, as well as flagellated rotary motors, which are driven by the proton gradient between the cytosol and the extracellular environment.

The working of the chemoreceptor cluster is best characterized by the analogy of an antenna. Similar to amplifying weak radio signals, receptor clusters allow the cell to amplify tiny changes in ligand concentration. This requires receptors to work not individually, but in a team. For this to be physically possible, receptors need to be coupled with their neighboring receptors so that binding of one ligand molecule affects a whole group of receptors (see Fig. 2.4). In addition to the receptors, a

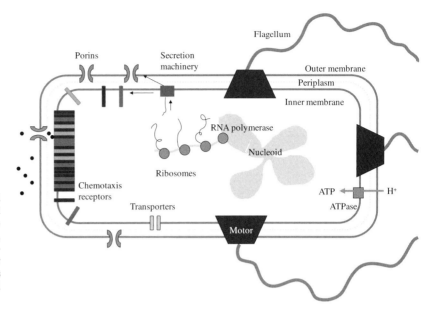

**Fig. 2.3** The biochemistry of bacterial chemotaxis. Receptors are made and inserted in the inner membrane. Chemicals pass the outer membrane via pores, are sensed in the periplasm, and transported through the inner membrane by transporters. Flagellated rotary motors are driven by a proton gradient and allow the cell to propel itself forward.

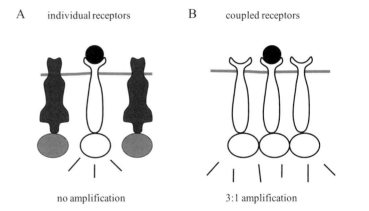

**Fig. 2.4** Signaling by individual receptors (A) and coupled receptors (B). The coupling of receptors leads to amplification and signal integration.

signal transduction pathway is required to relay the external chemical stimuli to the rotary motors, which drive the flagella for swimming.

## 2.3  *E. coli* chemotaxis pathway

The chemotaxis pathway of *E. coli* is shown in Fig. 2.5A. Sensing starts with the chemoreceptors, of which there are different types with different ligand specificities. Imagine that ligand molecules have different shapes, which fit perfectly into the ligand-binding pocket of the receptor in a lock-and-key fashion (see Fig. 2.2A). The most abundant receptor types are the Tar and Tsr receptors, which bind the amino acids aspartate and serine, respectively. There are also three other, less abundant receptor types. Receptors in the active conformation are

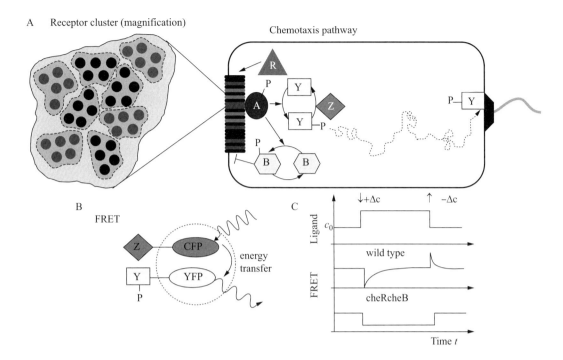

**Fig. 2.5** (A) Schematic of the chemotaxis pathway in *E. coli* (top right) with a magnified receptor cluster with smaller receptor complexes (top left). Receptors, kinase CheA, phosphatase CheZ, response regulator CheY, and adaptation enzymes CheR and CheB are shown. CheR adds methyl groups to receptors, and CheB, once phosphorylated, removes methyl groups. (Phosphate groups are indicated with the letter P.) Specifically, CheY-P can, by diffusion (dashed trajectory), find and subsequently bind motor to induce a switch from running to tumbling. (B) Illustration of FRET measurement. Donor fluorophore CFP is attached to CheZ, and acceptor fluorophore is attached to YFP. Once in proximity, energy transfer can occur. (C) The FRET response of adapting (wild type) and non-adapting mutant cells (with enzymes CheR and CheB removed) are shown. (B,C) Redrawn from Ref. (7).

able to activate the receptor-bound molecule CheA to phosphorylate itself—a so-called kinase—by using the fuel molecule ATP. It then passes the phosphate group onto the two other molecules, CheY and CheB—so-called response regulators. Phosphorylated CheY, i.e. CheY-P, diffuses through the cytoplasm and binds to multiple motors (specifically the FliM motor protein), inducing their clockwise (CW) rotation. The flagella fly apart incoherently, which leads to tumbling and random reorientation of the cell. This occurs more frequently when the cell happens to swim in the wrong direction. If the receptors are in the inactive conformation, the kinase does not autophosphoryate. Consequently, CheY is not phosphorylated, since another molecule CheZ—a so-called phosphatase—turns CheY-P back into CheY, which does not bind to the motors. Hence, the motors are in their default mode of counter-clockwise (CCW) rotation. This leads to bundling of the flagella, allowing the cell to run straight. This occurs more frequently when the cell happens to swim in the right direction.

In order for cells to be able to sense new changes in chemical concentration, they need to adapt. Adapted cells always have the same probability of tumbling, irrespective of the external chemical concentration.

Adaptation happens at the receptors. For this purpose, enzyme CheR covalently attaches methyl groups to specific glutamate amine acids at the cytoplasmic (cell internal) part of the receptors using fuel molecule SAM. Unlike the negatively charged glutamates, methylated glutamates are charge-neutral and hydrophobic. Such receptor methylation increases the signaling activity and the cell's probability of tumbling. There is also enzyme CheB, which, when phosphorylated by activated CheA (i.e. CheB-P), demethylates the receptors, and turns methylated glutamates back into glutamates. This decreases the signaling activity and increases the cell's probability of running. The charge state and degree of hydrophobicity of the receptors is believed to influence the equilibrium conformation of the receptors, thus allowing the external signal to transverse the inner cell membrane.

## 2.4   Experimental approaches

To study bacterial chemotaxis by light-microscopy, researchers tracked swimming bacteria to record their runs and tumbles, or tethered cells with their motor or flagella to a glass slide to observe their direction of rotation. While these are powerful techniques, signaling can also be measured directly in living cells, using fluorescence resonance energy transfer (FRET). For this purpose, a donor fluorophore needs to be attached to one type of molecule in the pathway, and an acceptor fluorophore to another type of molecule, as shown in Fig. 2.5B. These fluorophores are genetically attached to the proteins, and need to have specific optical properties. The donor fluorophore is excited by a laser at some wavelength, and after excitation, emits light at a longer wavelength. However, when the acceptor comes close to the donor (due to interaction in the signaling pathway), the acceptor can absorb the excitation energy, and emit the light in a different, longer wavelength. In *E. coli*, CheZ is genetically fused to donor CFP, and CheY is fused to acceptor YFP, which are so-called fluorescent proteins. The ratio between the two fluorescence intensities is a measure of the signaling activity in real time.

Using FRET, signaling and adaptation can be quantified thoroughly, greatly facilitating the study of the remarkable sensing and signaling properties outlined earlier. Using stimulation by changes in attractant concentration, the response and subsequent adaptation can be studied. Particularly convenient is attractant MeAsp, which is a non-metabolizable analog of aspartate and inhibitor of the Tar receptor. As an example, schematic Fig. 2.5C shows cells initially adapted to a background concentration of MeAsp and subsequently exposed to a step increase of attractant as applied in an experimentally controllable flow chamber. After adaptation, the step is removed and cells adapt again to the original background concentration.

Recording the initial responses in activity (from FRET) results in dose-response curves, which provide detailed information about how

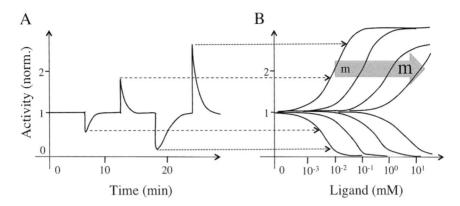

**Fig. 2.6** (A) Time course of activity as can be measured by FRET in response to increasingly higher step additions and removals in attractant MeAsp, illustrated for zero background concentration. (B) Corresponding dose-response curves of initial response amplitudes for this and three additional background concentrations. Due to adaptation to different background concentrations, the receptor methylation level $m$ increases from left to right, as indicated by the arrow. The activity is normalized by the adapted value. Redrawn from Ref. (7).

inputs (concentration) are mapped onto outputs (activity from FRET). This is illustrated in Fig. 2.6, in which the time-dependent responses to increasingly higher step additions and removals (panel A) lead to a dose-response curve (panel B). Also shown are three more dose-response curves corresponding to higher background concentrations. Inspection of these dose-response curves shows that adaptation changes the ligand concentration corresponding to half-maximal activity. When adapted to zero background concentration, receptors exhibit an extremely high sensitivity to addition of attractant (very left dose-response curve). At larger and larger background concentrations, receptors become more and more insensitive in responding to changes in attractant concentration, presumably due to the increase in the receptor methylation level. Hence, methylation effectively desensitizes receptors.

## 2.5   Time-course data and dose-response curves

The bacterial chemotaxis proteins are easily modified by genetics, allowing researchers to make cells with specific properties. For instance, cells can be made which express only a single receptor type. Furthermore, they can be made which express the receptor in specific fixed modification (methylation-like) states. The making of modification states is possible because a receptor with a glutamine amino acid behaves similarly to a receptor with a methylated glutamate; both are hydrophobic and affect the receptor conformation similarly. Enzymes CheR and CheB also need to be removed from the chromosome (naturally occuring DNA) so that the remaining glutamates are not methylated. These receptor modifications can be made either in the chromosome or by using a plasmid, which is a small extra piece of DNA that encodes additional proteins.

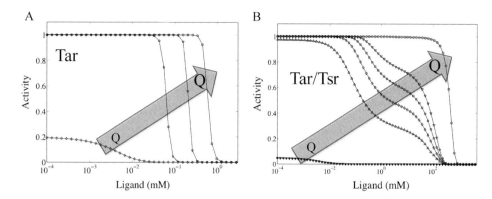

**Fig. 2.7** Dose-response curves of activity as can be measured by FRET of non-adapting cells with receptors in fixed, genetically engineered modification states with stimulation by attractant MeAsp. Cell types either express only the Tar receptor (A) or mixed receptor types, mostly Tar and Tsr (B). In both cases, the modification level (amount of glutamines Q) of the Tar receptor increases from the left to right curves in each panel, as indicated by arrows. Panels (A) and (B) based on Refs. (10) and (7), respectively.

Fig. 2.5C shows, in addition to the adapting "wild type", the response of mutant cells with receptors in a fixed modification state. Unlike the adapting cells, the mutant cells simply show a reduced activity in the presence of the attractant. Once the attractant is removed, the activity returns to the original activity. This demonstrates that these genetically altered cells behave in an expected, controllable way.

To reduce the signaling complexity even further, Fig. 2.7A shows dose-response curves of cells expressing only Tar receptors in different modification states, with the number of glutamines per receptor increasing from left to right. These experiments show that unmodified receptors are highly sensitive as they turn off their activity at small amounts of attractant, while there are large Hill coefficients and hence cooperativity for high modification levels. Both these effects point towards strong receptor–receptor coupling. In contrast, Fig. 2.7B shows dose-response curves of cells expressing all receptors, with Tar and Tsr the most abundant receptor types. This clearly shows the increased dynamic range of the receptor activity, now spanning up to five orders of magnitude in attractant concentration. Also apparent is the biphasic response characteristics in the presence of multiple receptor types. These features certainly beg for quantitative explanation.

Before moving on to the theoretical explanations of how the chemotaxis sensory apparatus and pathway work, let us discuss a few final aspects of the dose-response curves shown so far. Fig. 2.6B shows the change in activity $\Delta A$ of initially adapted receptors for different changes in attractant $\Delta c$. In contrast, Fig. 2.7 shows the activity $A$ as a function of attractant concentration $c$. To ensure that we understand the difference between the two types of curves, we illustrate their relatedness in Fig. 2.8. In this figure, the dose-response curve in the middle corresponds to receptors adapted to ligand concentration $c_0$, methylation level $m$, and adapted activity $\bar{A}$. A sudden addition of attractant $(+\Delta c)$ leads to an almost instantaneous decrease in activity for the same

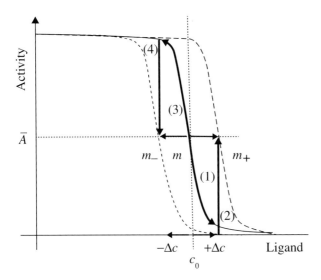

**Fig. 2.8** Illustration of adaptation with three dose-response curves, corresponding to three different methylation levels (or three different attractant concentrations). Initially, receptors are adapted to concentration $c_0$ with methylation level $m$. When attractant is added with new concentration $c_0 + \Delta c$, receptor activity initially drops with subsequent adaptation to the new methylation level $m_+ > m$. Instead, when attractant is removed with new concentration $c_0 - \Delta c$, receptor activity initially increases with subsequent adaptation to the new methylation level $m_- < m$.

methylation level following arrow (1). Subsequently, the receptor activity adapts back to $\bar{A}$ following arrow (2). The corresponding methylation level is $m_+$ with the associated dose-response curve shown with a long dashed line. Alternatively, removal of attractant concentration $(-\Delta c)$ leads to an increase in activity along arrow (3), followed by adaptation along arrow (4). This leads to the dose-response curve plotted with the short dashed line and methylation level $m_-$. Hence, we can plot dose-response curves either as a function of changes in ligand concentration relative to the adapted activity (Fig. 2.6B), or as a function of total ligand concentration (Fig. 2.7).

# Chapter summary

- Chemotaxis in *E. coli* shows remarkable sensing and signaling properties, including high sensitivity in detecting minute changes in chemicals such as attractants, precise adaptation to persistent changes, and a wide dynamic range over which chemotaxis works. Since the pathway is very well characterized, *E. coli* chemotaxis is a paradigm for sensing and signaling, ideal for quantitative modeling.

- To measure the chemotactic response of cells, researchers can do the following: track swimming cells, tether cells to a substrate to study individual motors, observe populations of cells in microfluidic chambers of controlled gradients, or use *in vivo* FRET to study signaling of immobilized cells in a flow chamber of controlled chemical concentrations.

- *In vivo* FRET allows the measurement of time courses of response and subsequent adaptation with excellent signal-to-noise ratio.

The initial response amplitudes belonging to the same background concentration prior to stimulus can be combined in the form of a dose-response curve. Such curves can be measured for adapting cells, as well as for non-adapting mutant cells.

# Further reading

(1) '*E. coli in Motion*', by Howard C. Berg (Springer, 2004).

(2) '*Sensory Transduction*', by Gordon L. Fain (Sinauer, 2003).

(3) '*Cell Movements: From Molecules to Motility*', by Dennis Bray (Garland, 2000).

(4) 'Chemotaxis in bacteria', by Julius Adler, *Science* 153: 708–16 (1966).

(5) 'Chemotaxis in *Escherichia coli* analysed by three-dimensional tracking', by Howard C. Berg and Douglas A. Brown, *Nature* 239: 500–04 (1972).

(6) 'Amplification and adaptation in regulatory and sensory systems', by Daniel E. Koshland Jr, Albert Goldbeter, and Jeffry B. Stock, *Science* 217: 220–5 (1982).

(7) 'Receptor sensitivity in bacterial chemotaxis', by Victor Sourjik and Howard C. Berg, *Proc. Natl. Acad. Sci. USA* 99: 123–7 (2002).

(8) 'Functional interactions between receptors in bacterial chemotaxis', by Victor Sourjik and Howard C. Berg, *Nature* 428: 437–41 (2004).

(9) 'Receptor clustering and signal processing in *E. coli* chemotaxis', by Victor Sourjik, *Trends Microbiol.* 12: 569–76 (2004).

(10) 'Variable sizes of Escherichia coli chemoreceptor signaling teams', by Robert Endres, Olga Oleksiuk, Clinton H. Hansen, Yigal Meir, Victor Sourjik, and Ned S. Wingreen, *Mol. syst. Biol.* 4: 211 (2008).

# Physical concepts

3

Before addressing actual problems in chemotaxis, we first review a number of important physical concepts. These will serve as a foundation for understanding the material in the following chapters.

Physics can be considered the most general approach to understanding both non-living and living matter. It aims to explain universal aspects, often robust to underlying details, and can be very useful in understanding constraints and optimality in physical systems. Although biology has added many layers on top of physical mechanisms, it appears to tinker with what physics proposes naturally—not fighting physics, but regulating it. In this chapter we will discuss a range of topics, including diffusion, statistical mechanics as applied to molecular binding events, and the role of thermal fluctuations, which are always present in biological systems. Additional material can be found in Appendix A.

## 3.1 Diffusion

Diffusion, or Brownian motion, as named after its discoverer, describes the random movement of molecules in an aqueous solution in thermal equilibrium (see Fig. 3.1). Thermal equilibrium does not mean that a molecule comes to rest as it would in our macroscopic world, but rather, it keeps juggling around through the constant bombardment by surrounding water molecules. Historically, the theory of diffusion was also an indirect confirmation of the atomistic nature of matter. In biology, diffusion is the most common means of transport in a cell and hence crucial for signaling. Thus, it is important to gain good intuition of how diffusion works, and hence diffusion will be covered in some detail. Specifically, we will study the effect of the spatial dimension on diffusion. Furthermore, since modeling of diffusion[1] can historically be considered one of the first attempts to model a biological process, we will discuss the various theoretical approaches and their limitations. This is a great example of the fact that even "wrong" models can be useful in making predictions; in light of discrepancies with experiments, such erroneous models need revision, which subsequently leads to improved but generally more complicated models.

[1] Most noticeably are the approaches by Albert Einstein (1905), Marian Smoluchowski (1906), Paul Langevin (1908), and Norbert Wiener (1923), although there have been attempts dating back as early as 60 BC.

### 3.1.1 Random walk

The simplest way of modeling diffusion is by implementing a basic algorithm for random walk on a lattice on which a particle randomly

**Fig. 3.1** Diffusing particle in 2D space, produced by the algorithm described in the text.

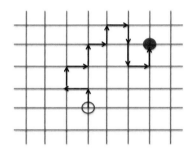

**Fig. 3.2** 2D lattice for simulating the random walk of a particle. At each time point of the simulation, the particle "throws a dice" to decide in which direction to make the next step.

moves from site to site. For simplicity we consider a one-dimensional (1D) lattice for the derivation, but display a 2D lattice in Fig. 3.2 for illustrative reasons. Although a 1D model appears highly idealized, this could represent a gene-regulatory protein moving along DNA due to non-specific attraction between the oppositely charged molecules. 1D diffusion could also be used to describe bacterial movement in a long thin glass tube.

To set up the algorithm in 1D, assume a particle can step at each time point either left or right by an amount $\Delta x$. The position $x_n$ of the particle starting at $x_0 = 0$ after $n$ steps is

$$x_n = \Delta x_n + \Delta x_{n-1} + \cdots + \Delta x_1 = \sum_{i=1}^{n} \Delta x_i. \tag{3.1}$$

After squaring and taking the average over a large number of particles, we obtain the mean-squared path length

$$\langle x_n^2 \rangle = \sum_{i=1}^{n} \langle \Delta x_i^2 \rangle + \sum_{i,j=1 (i \neq j)}^{n} \langle \Delta x_i \Delta x_j \rangle, \tag{3.2}$$

where the sum of the second term excludes indices for which $i$ and $j$ are equal.

In Eq. (3.2), the first term on the right-hand side is simply $n\Delta x^2$ and, since $\Delta x_i$ and $\Delta x_j$ are independent and their individual averages are zero, the second term is zero. Now we assume that the total time elapsed after $n$ moves is $t = n\Delta t$, where $\Delta t$ is the time for each left or right move. We can use this to write the path length in terms of the time

$$\langle x(t)^2 \rangle = \frac{\Delta x^2}{\Delta t} t = 2Dt, \tag{3.3}$$

where we introduced the diffusion constant $D = \Delta x^2/(2\Delta t)$. For convenience this includes a factor 2, which becomes more apparent when alternatively deriving diffusion using Fick's laws of particle currents.[2] The root-mean square (rms) displacement has the characteristic square-root time dependence

$$\text{rms} = \langle x(t)^2 \rangle^{1/2} = (2Dt)^{1/2}. \tag{3.4}$$

In the random-walk algorithm, the rules of random left or right movement can be applied to any dimension with the displacement given by $r^2 = x^2 + y^2 + \dots$. As each dimension is independent from the others, the path length becomes $4Dt$ in two dimensions and $6Dt$ in three dimensions, and so on.

I make a final remark on the mean displacement. In Eq. (3.2), if we had assumed $\Delta x_i = \Delta x_j = v\Delta t$ for directed, ballistic movement with velocity $v$, then the double sum would have produced $n(n-1) \approx n^2$ for $n \gg 1$. This leads to

$$\langle x(t)^2 \rangle = 2Dt + (vt)^2 \tag{3.5}$$

instead, and hence, for long times, produces a quadratic time dependence.

### 3.1.2  Properties of diffusion

When pondering diffusion in different dimensions the reader may wonder about three things. First, how effective is diffusion to transport molecules in cells over distances such as from cell end to cell end? Second, if a molecule is searching for its reaction partner by diffusion, how thorough is this search? Is the particle searching the space evenly or does it easily miss parts? Third, if two molecules meet, how likely is it that they meet again if they diffused apart before reacting?

Diffusion works well over small distances, but is an ineffective way of transport over long distances. To demonstrate this, we use the diffusion constant $D = 100\,\mu\text{m}^2/\text{s}$ of a small molecule in water. To travel a micrometer, approximately corresponding to the size of a bacterium, takes only about $t = x^2/(2D) = 5$ ms. In contrast, to travel a distance $x = 30\,\mu m$ corresponding to the dimension of a larger eukaryotic cell, takes already a significant larger amount of time $t \approx 5$ s, which is too slow for many cellular processes.[3] This strong distance dependence of the diffusion time is a direct effect of the fact that the time is not directly proportional to the distance but the square of it. Hence, transport by motor proteins along protein filaments or by electrical excitations of the membrane, e.g. along neural axons of our nervous system, are faster ways at the expense of higher energy consumption. (Remember that diffusion is for "free".)

Diffusion is a thorough search in 1D and 2D, but not in higher dimensions. To gain an intuitive understanding, we use the *central limit theorem*[4], i.e. that each step of a random walk is itself a random variable, and so the sum of them is another random variable with a Gaussian

[2] Adolf Fick (1855).

[3] Biological cells range from $1\mu$m to $100\mu$m. Eukaryotic (nucleated) cells are typically ten times the size of prokaryotic (bacterial) cells.

[4] This proves use again and again.

distribution (zero average and finite variance). Consider a lattice in $d$ dimensions. The particle visits a sequence of $N$ sites, effectively confined to the "central region" of a volume of order $O(N^{d/2})$, or linear dimension $O(\sqrt{N})$. (As $N \to \infty$, it is exponentially unlikely to find the particle outside this region.) For $d = 1$, the number of visited sites $N$ grows faster than the number of accessible sites $\sqrt{N}$, so the particle's trajectory must loop back and forth frequently, returning to the origin (and other visited sites) many times as $N \to \infty$. On the other hand, for $d \geq 3$, the number of accessible sites in the central region grows faster than the number of visited sites, so the particle will explore new sites without necessarily returning to any previously visited site. Diffusion in 2D is a borderline case and needs more detailed analysis. Interestingly, the random walk of a particle creates a fractal[5] of dimension 2 when embedded in higher dimensions.

[5]A rough or fragmented geometric shape with self-similarity, i.e. that each part is a reduced-size copy of the whole.

It turns out that particles will meet again in one or two dimensions (probability equal to 1) but not necessarily in three and higher dimensions (probability smaller than 1). This is known as Polyá's theorem, and can be easily demonstrated in 1D: consider a particle performing random walk on the 1D lattice (a sequence of integers starting at the origin, i.e. at zero). We would like to determine the probability $P$ that a particle eventually returns to 0 regardless of the number of steps it takes. Without loss of generality, assume the first step is to the right, to 1. Now the particle can either go left with probability $1/2$ or can return to 1 a number $n$ times before returning to 0 with probability $1/2$. Since half of the trajectories from 1 back to 1 remain on the right side of 1, the probability of returning to 1 just once before going back to 0 is $P/2$. Similarly, the probability of returning to 1 $n$ times before going to 0 is $(P/2)^n$. As a result,

$$P = \frac{1}{2} + \left[\sum_{n=1}^{\infty}(P/2)^n\right]\cdot\frac{1}{2} = \frac{1}{2}\left[\sum_{n=0}^{\infty}(P/2)^n\right] = \frac{1}{2(1 - P/2)} = \frac{1}{2 - P},$$
(3.6)

which is solved by $P = 1$. Hence, a particle performing random walk in 1D always returns back to its origin, although this can take a long time. In fact, the expected return time is infinite. The result is intuitively clear—the chance to find a specific location is greatly reduced in higher dimensions. This is the curse of dimension, and not only affects molecular reactions but computational search algorithms. It is the reason that chemical reactions most probable occur at filaments and membranes, which can be considered low-dimensional reactive surfaces.

### 3.1.3   Diffusion equation

The random walk algorithm regards diffusion from the perspective of a single particle. However, there is also a macroscopic perspective on diffusion based on concentrations, densities, or probability distributions. Diffusion is macroscopically visible if there is a spatial difference in concentration and the region of the high concentration spreads out into the region of low concentration. We also know that particles are not made or removed during the process of diffusion. Using these observations, we can consider diffusion along the $x$-direction in 3D space, and imagine two areas perpendicular to the flow (in the $y-z$ plane) a distance $\Delta x$ apart (see Fig. 3.3). Particle conservation implies that the time variation of the concentration $c(x,t)$ inside elementary volume $\Delta x \Delta y \Delta z$ equals the influx minus the outflow of particles. Hence, using the particle flux $J(x,t)$, we can write

$$\frac{\partial c(x,t)}{\partial t}\Delta x \Delta y \Delta z = [J(x) - J(x + \Delta x)]\Delta y \Delta z = -\frac{\partial J}{\partial x}\Delta x \Delta y \Delta z, \quad (3.7)$$

which leads to the diffusion equation in its general form

$$\frac{\partial c(x,t)}{\partial t} = -\frac{\partial J}{\partial x}. \quad (3.8)$$

Now, all we need to do is determine the flux $J$. From its physical definition, it is obvious that

$$J(x,t) = c(x,t)v(x,t) \quad (3.9)$$

with $v$ the average velocity, which we do not know. However, we use the empirical observation[6] that the flux across an area in the $y-z$ plane is proportional to the concentration gradient along the $x$ direction

$$J_x = -D(x)\frac{\partial c}{\partial x}, \quad (3.10)$$

where $D$ is the diffusion constant, which may depend on spatial coordinates, e.g. due to local crowding effects or available binding sites for the particles. With this expression for the flux, the diffusion equation takes the form[7]

$$\frac{\partial c(x,t)}{\partial t} = \frac{\partial}{\partial x}\left[D(x)\frac{\partial c(x,t)}{\partial x}\right] \quad (3.11)$$

or, for constant diffusion constant $D$,

$$\frac{\partial c(x,t)}{\partial t} = D\frac{\partial^2 c(x,t)}{\partial x^2}. \quad (3.12)$$

For infinite space, and all particles concentrated at $x = 0$ initially, the solution is a Gaussian distribution

$$c(x,t) = \frac{c_0}{\sqrt{4\pi D t}}e^{-x^2/(4Dt)}, \quad (3.13)$$

shown in Fig. 3.4, where $c_0$ equals the *total number* of particles over the whole volume. By inspecting the exponential, we obtain the same variance as before from our random walk algorithm, $\langle x^2(t)\rangle = 2Dt$. This relation is key for determining experimental diffusion constants.

The models of diffusion so far consider only the position of the particles, but ignore its momentum or velocity. This is a rather strong model simplification, and leads to the unphysical result that the average particle velocity of the particles in 1D behaves as

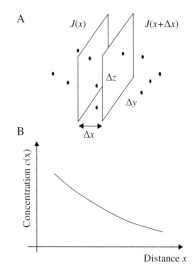

**Fig. 3.3** Flux $J$ in $x$-direction through two parallel planes in $y$-$z$ plane (A) in concentration profile (B).

[6]Fick's first law.

[7]Fick's second law.

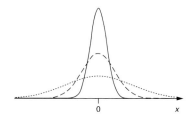

**Fig. 3.4** Particle distribution, which spreads out with time due to diffusion (solid line, dashed line, and dotted line for increasing times).

$$\frac{d\sqrt{\langle x^2(t)\rangle}}{dt} = \sqrt{\frac{D}{t}}. \tag{3.14}$$

[8]This is a problem with Einstein's treatment of diffusion.

For small times, this velocity diverges,[8] which certainly cannot be true, and thus we require an improved mathematical treatment.

### 3.1.4   Langevin approach

To remedy the problem of the diverging velocity, we treat both position and velocity of a particle in one coherent framework. We start by writing down the following differential equation (written for simplicity in 1D)

$$m\ddot{x} = -\gamma\dot{x} + f(t), \tag{3.15}$$

where we introduced the short notations $\dot{x} = dx/dt$ and $\ddot{x} = d^2x/dt^2$ to simplify expressions. Eq. (3.15) is stochastic due to the random force $f(t)$. This force acts on the particle due to collisions with the surrounding water molecules. Furthermore, $\gamma\dot{x}$ is the friction force, proportional to the velocity, with $\gamma$ the damping rate constant. Without the random force, this term would slow down an initially moving particle and make it come to rest.

To solve the *Langevin equation* (3.15), in particular to obtain correlation functions (see below), we multiply by $x$ and obtain

$$mx\ddot{x} = m\left[\frac{d(x\dot{x})}{dt} - \dot{x}^2\right] = -\gamma x\dot{x} + xf(t). \tag{3.16}$$

We now take the average and assume $\langle xf(t)\rangle = 0$ due to the irregular nature of the force, as it is caused by many water molecules,

$$m\frac{d\langle x\dot{x}\rangle}{dt} = m\langle \dot{x}^2\rangle - \gamma\langle x\dot{x}\rangle. \tag{3.17}$$

[9]In classical statistical mechanics, the equipartition theorem is a useful formula that relates the temperature of a system to its average energies, applicable to all degrees of freedom.

The second term on the right-hand side is just two times the kinetic energy of the particle. According to the *equipartition theorem*,[9] we have $m\langle\dot{x}^2\rangle/2 = k_BT/2$, where $k_B$ is the Boltzmann constant and $T$ is the absolute temperature in units of Kelvin, and we obtain

$$\left(\frac{d}{dt} + \frac{\gamma}{m}\right)\langle x\dot{x}\rangle = \frac{k_BT}{m}, \tag{3.18}$$

[10]Separation of variables is one of several methods for solving differential equations by rewriting an equation so that each of two variables occurs on a different side of the equation.

which can be solved with the technique called *separation of variables*[10] (see Chapter 4). This procedure returns

$$\langle x\dot{x}\rangle = \frac{1}{2}\frac{d\langle x^2\rangle}{dt} = Ce^{-\gamma t/m} + \frac{k_BT}{\gamma}. \tag{3.19}$$

Here, parameter $C$ is a constant of integration and needs to be determined from the initial condition that at $t = 0$ the mean-square displacement is zero, which leads to $C = -k_BT/\gamma$. Upon integration, we finally obtain

$$\langle x^2 \rangle = \frac{2k_BT}{\gamma}\left[t - \frac{m}{\gamma}(1 - e^{-\gamma t/m})\right]. \qquad (3.20)$$

For times much shorter than the collision time, i.e. in the limit $t < < m/\gamma = \tau_c$, the exponential function in Eq. (3.20) can be Taylor-expanded,[11] $e^{-t/\tau_c} \approx 1 - t/\tau_c$, which leads to the approximate solution $\langle x^2 \rangle \approx t^2$. This is called ballistic movement and corresponds to fast movement without hindrance by collisions. For large times, i.e. $t \gg \tau_c$, the solution has the familiar form $\langle x^2 \rangle \approx (2k_BT/\gamma)t = 2Dt$, where we applied the *Einstein–Smoluchowski* relation $D = k_BT/\gamma$ to introduce the diffusion constant. Hence, we see that a more sophisticated model based on particle position and velocity results in an improved physical description of the diffusion process. However, less correct models were key in arriving at this stage.

[11]In mathematics, a Taylor series expresses a function as an infinite sum of terms that are calculated from the values of the function's derivatives at a single point. Put simply, it tries to approximate a function locally as a low-order polynomial.

## 3.2   Boltzmann distribution

Imagine the sedimentation of small particles in solution. Initially the particles will move downward, driven by the lowering of their gravitational energy. However, if we wait long enough, not all particles will be at the very bottom. There will be a distribution of particles with fewer and fewer particles in higher fluid layers, as illustrated in Fig. 3.5. The distribution is named after *Boltzmann*[12] and is given by

[12]Austrian physicist Ludwig Boltzmann (1844–1906) is considered the god-father of statistical mechanics.

$$p_i = \frac{e^{-\frac{E_i}{k_BT}}}{Z}, \qquad (3.21)$$

where we assumed for simplicity discrete energy levels $E_i$ with index $i$ ($i = 1, \ldots, \Omega$, here for a finite number of states). These correspond, via the potential energy $E_i = mgx_i$, to height $x_i$ above the ground with $m$ the particle mass and $g$ the gravitational constant. The denominator is the *partition function* and includes all possible heights and energy levels. It is written as $Z = \sum_j e^{-E_j/(k_BT)}$, that is as the sum of all possible Boltzmann factors, $\exp[-E_j/(k_BT)]$. (For continuous energy levels, sums can to be turned into integrals.) While this example is about sedimentation of particles, the formalism is very general and can describe any equilibrium distribution as long as particles can be distributed over energy levels. For most cases, the relevant energy scale is the thermal energy $k_BT$.

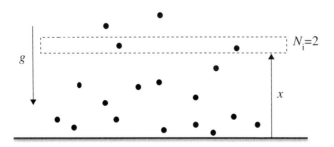

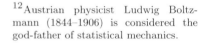

**Fig. 3.5** Sedimentation of particles in solution under the influence of the gravitational field. The potential energy of a particle is $mgx$ with mass $m$, gravitational force constant $g$, and height above the ground $x$. The dashed box indicates a discrete energy level, "occupied" by two particles. Redrawn from Ref. (2).

According to the equipartition theorem, half of the thermal energy is available to any degree of freedom of the particle, such as translational movement in height. Hence, the average potential energy is given by $\langle E \rangle = k_B T / 2$, but the same would apply to, e.g., the kinetic energy or the rotational energy. In fact, while particles are not static and instead constantly move in all directions due to diffusion, there is no net change in the particle distribution; any particle moving upwards is, on average, balanced by a particle moving downward. This condition of equilibrium is called *detailed balance*.[13] Since the Boltzmann distribution is of such fundamental importance, we briefly sketch one of its many derivations.

Assume we have $N$ particles with total energy $E$. Imagine that these particles are distributed with $N_i$ particles in energy level $E_i$ in such a way that constraints $N = \sum_i N_i$ and $E = \sum_i N_i E_i$ are fulfilled. Note by dividing by $N$, we can equivalently write $1 = \sum_i p_i$ and $\langle E \rangle = \sum_i p_i E_i$ with fraction $p_i = N_i / N$. How many ways can we realize this particular distribution of the particles? There are generally $N(N-1)(N-2)\ldots 2 = N!$ ways to distribute $N$ classical, distinguishable particles.[14] Since permutations of the particles on the same energy level do not change the individual particle's assignment, we can divide out these arrangements and obtain for the multiplicity

$$m = \frac{N!}{N_1! N_2! N_3! \ldots}. \tag{3.22}$$

Now the distribution with a large multiplicity $m$ ought to be more likely than a distribution with a very small multiplicity. This reflects maximizing the entropy of the system,[15] which can be formalized using Shannon's information entropy, introduced in 1948. In information theory, entropy is a measure of the uncertainty associated with a random variable. Using a natural logarithm we define (see Chapter 4)

$$H = -\sum_i p_i \ln p_i \tag{3.23}$$

with $H \geq 0$. For instance, if all probabilities are equal with $p_i = 1/\Omega$, the entropy takes on the maximal value $S = \ln \Omega$. If on the other hand, one of the probabilities is equal to 1 and the remaining ones are zero, i.e. we know exactly the state of the system, then $S = 0$ as there is no more ignorance. Returning to our particles in the gravitational field, we can write Shannon's entropy as

$$H' = -\sum_i p_i \ln p_i - \alpha \left[ \sum_i p_i - 1 \right] - \beta \left[ \sum_i p_i E_i - \langle E \rangle \right] \tag{3.24}$$

with which we extended to include the two constraints, that is normalization of the probability (or equivalently conservation of particles) and fixed average energy $\langle E \rangle$. These two constraints are imposed using *Lagrange multipliers* $\alpha$ and $\beta$. Maximizing entropy $H'$ by setting its derivative with respect to the unknown probabilities to zero, i.e. $\partial H'/\partial p_i = 0$, quickly leads to the solution

[13]The principle of detailed balance applies to kinetic systems with elementary processes such as collisions, or steps, or elementary reactions. It says that at equilibrium, each elementary process should be equilibrated by its reverse process.

[14]At low temperatures, quantum mechanical indistinguishable particles would behave very differently.

[15]There is also an analogue approach to non-equilibrium systems called *maximum caliber*.

$$p_i = \frac{e^{-\beta E_i}}{\sum_i e^{-\beta E_i}}. \qquad (3.25)$$

The remaining Lagrange multiplier is determined as $\beta = 1/(k_B T)$ by calculating the average energy $\langle E \rangle = k_B T/2$.

Note that the entropy in statistical mechanics is defined differently, but can be considered a special case of Shannon's general description. There is the Gibbs entropy

$$S = -k_B \sum_i p_i \ln p_i \qquad (3.26)$$

for ensembles of physical systems, and the Boltzmann entropy $S = k_b \ln \Omega$ for equilibrium systems, where each microstate is equally likely. The latter is, of course, equivalent to the Boltzmann distribution.

## 3.3   Ligand-receptor binding

We can now put the Boltzmann distribution into action and apply it to calculating the probability that a receptor is ligand-bound, the fundamental first step in largely any signaling cascade. (A very closely related problem is represented by protein-DNA binding for regulating gene expression.)

As shown in Fig. 3.6, we start by representing space by a lattice of $\Omega$ sites, which can be occupied by $L$ ligand molecules in solution.[16] Each ligand molecule in solution has an energy $\epsilon_s$. There is also a receptor present, which can bind at most one ligand molecule with the bound ligand energy $\epsilon_b$ assumed lower than $\epsilon_s$. Now, once in a while, a ligand

[16]Here we follow the nice treatment in Ref. (1).

A

Ligand-unbound receptor

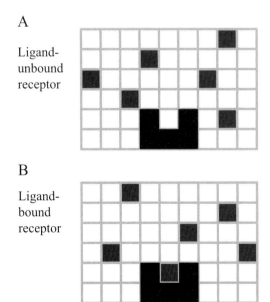

B

Ligand-bound receptor

**Fig. 3.6** Equilibrium properties of ligand-receptor binding can be calculated on a lattice. (A) Ligand is not bound to receptor. (B) Ligand is bound to receptor. Black object indicates receptor, gray squares represent ligand molecule. Redrawn from Ref. (1).

molecule will bind, lowering the total energy of ligands and the receptor by the binding energy $\Delta\epsilon = \epsilon_b - \epsilon_s$. Such binding also reduces the number of free ligand molecules by one, leaving $L - 1$ in solution. Subsequently, the ligand molecule may unbind again, mixing itself among the other ligand molecules. It is rather difficult to keep track of all the binding and unbinding events. However, to calculate the equilibrium probability of the receptor being bound, all we need are energies (and no kinetics). Entropy is also important, which is related to the number of microstates, i.e. the arrangements of molecules on the lattice, with the receptor either bound or unbound.

To calculate the probability of the receptor to be bound by a ligand molecule, we need to calculate the ratio of weights of unbound over either bound or unbound

$$p_b = \underbrace{\frac{\text{weight(bound)}}{\text{weight(unbound)} + \text{weight(bound)}}}_{\text{partition function}}. \qquad (3.27)$$

Specifically, the weight of the bound receptor is the product of the Boltzmann factor for the bound state, $\exp\{-[\epsilon_b + (L-1)\epsilon_s]/(k_B T)\}$, and the multiplicity $m_b$ of this state. How many microstates are there? There are $L - 1$ indistinguishable ligand molecules and $\Omega$ places to put them. Hence, there are $\Omega(\Omega - 1)(\Omega - 2)\ldots[\Omega - (L-2)] = \Omega!/(\Omega - (L-1))!$ ways of distributing the molecules, and since one ligand molecule cannot be told apart from another, their intrinsic order does not matter. Hence, we also divide this number by the number of ways of rearrangement, $(L-1)!$, and obtain for the multiplicity

$$m_b = \frac{\Omega!}{(L-1)!(\Omega - L + 1)!} \approx \frac{\Omega^{L-1}}{(L-1)!} \qquad (3.28)$$

where we used the *Stirling approximation*[17] valid for $\Omega \gg L$.

In the denominator, the sum of the weights for unbound and bound configurations is the partition function, which contains all possible states of the system with Boltzmann factors and corresponding multiplicities. Specifically, the energy of the unbound state is $L\epsilon_s$ with multiplicity

$$m_u = \frac{\Omega!}{L!(\Omega - L)!}. \qquad (3.29)$$

Taken together, the probability of being bound is given by

$$p_b = \frac{e^{-\frac{\epsilon_b}{k_B T}} \frac{\Omega^{L-1}}{(L-1)!} e^{-\frac{(L-1)\epsilon_s}{k_B T}}}{\frac{\Omega^L}{L!} e^{-\frac{L\epsilon_s}{k_B T}} + e^{-\frac{\epsilon_b}{k_B T}} \frac{\Omega^{L-1}}{(L-1)!} e^{-\frac{(L-1)\epsilon_s}{k_B T}}}. \qquad (3.30)$$

To simplify this expression further, we divide the numerator and denominator by $(\Omega^L/L!)\exp[-L\epsilon_s/(k_B T)]$ and obtain

$$p_b = \frac{\frac{L}{\Omega} e^{-\frac{\Delta\epsilon}{k_B T}}}{1 + \frac{L}{\Omega} e^{-\frac{\Delta\epsilon}{k_B T}}}, \qquad (3.31)$$

[17]In mathematics, the Stirling approximation applies to large factorials based on $\ln n! \approx n \ln n - n$ for large $n$.

where we used the binding energy $\Delta\epsilon$. Now we introduce the ligand concentration $c = L/V$ with $V$ the real volume. This is done by writing $L/\Omega = L/V \cdot V/\Omega = c/c_0$ with $c_0$ a reference ligand concentration, corresponding to all lattice sites $\Omega$ occupied with a ligand molecule.

We can now write Eq. (3.31) in two different ways. For instance, instead of using multiplicities and energies, we can introduce free energies by including the multiplicity terms in the exponentials

$$p_b = \frac{e^{-\frac{\Delta\epsilon - k_B T \ln(c/c_0)}{k_B T}}}{1 + e^{-\frac{\Delta\epsilon - k_B T \ln(c/c_0)}{k_B T}}}. \tag{3.32}$$

Now all weights are normalized by the weight of the unbound state and hence are written relative to this state. From this notation it becomes clear that there are two effects determining whether a ligand binds to the receptor. There is the energy gain $\Delta\epsilon$, the familiar binding energy, and the entropy cost (entropy benefit) from the ligand taken out of solution for $c < c_0$ ($c > c_0$). There is no net gain if both balance each other, $\Delta\epsilon = k_B T \ln(c/c_0)$. In this case, the receptor is consequently bound with 50% probability.

Alternatively, we can introduce the parameter $K_D = c_0 e^{\Delta\epsilon/(k_B T)}$, which is the smaller the stronger the binding (the more negative $\Delta\epsilon$), and write more compactly

$$p_b = \frac{c}{c + K_D}. \tag{3.33}$$

This notation makes clear that $K_D$ is the ligand concentration at which the probability of being bound is 50%. The parameter $K_D$ is called the *ligand-dissociation constant*.[18] If there is cooperative ligand binding to the receptor, more than one ligand can bind the same receptor and the binding of one ligand molecule makes it more favorable to bind additional ligand molecules. For all-or-none binding of ligands, the following *Hill equation* can be written

$$p_b = \frac{c^n}{c^n + K_D^n} \tag{3.34}$$

with $n \geq 1$ the *Hill coefficient* as a measure of cooperativity.[19] Plotting this equation produces a sigmoidal shape for cooperative binding (Fig. 3.7). We will will come back to cooperative ligand binding in Chapter 5.

[18]The ligand dissociation constant $K_D = K_{eq}^{-1}$ is the inverse of the equilibrium constant $K_{eq}$.

[19]The Hill coefficient was originally used to explain the cooperative binding of oxygen to hemoglobin (Hill coefficient of 2.8–3).

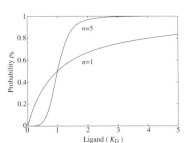

**Fig. 3.7** Plot of Hill function without ($n = 1$) and with ($n = 5$) cooperativity. Ligand concentration is in units of $K_D$.

## 3.4 Fluctuation–dissipation theorem

The *fluctuation–dissipation theorem* (FDT) of statistical mechanics is a very useful concept as it allows the calculation of the equilibrium fluctuations of a physical system in terms of the system's non-equilibrium response to a small perturbation. The latter can conveniently be calculated using *linear response theory*.[20]

[20]For a small stimulus or perturbation, linear response theory is a convenient way to calculate the response of a system, utilizing linearization of its components as a simplifying concept.

Imagine a protein bound to a molecular tether, such as the short unfolded polypeptide at the end of a receptor shown in Fig. 3.8. (We will see later that in bacterial chemotaxis, enzymes that modify receptors during adaptation do indeed tether to the receptors.) Let us measure the protein's horizontal distance from the receptor by observable $x$. In one conformation of the receptor, the enzyme is not bound to the main part of the receptor (but remains bound to the tether), and hence variable $x$ will just fluctuate around a value determined by energy $E_0(x)$. (For convenience, we use $\langle x \rangle_0 = 0$ for the unbound configuration.) Now at time $t = 0$ the receptor changes conformation, e.g. due to external binding of a ligand molecule, allowing the enzyme to bind the receptor. This new conformation is represented by an additional attractive force $f(t) < 0$, which might be time dependent and can modify the energy according to $E(x, t) = E_0(x) + x f(t)$. The response of the observable $x$ to the new force is characterized (to first order) by the susceptibility, or linear response function, $\chi(t)$ of the system

$$\langle x(t) \rangle = \langle x \rangle_0 + \int_{-\infty}^{t} f(\tau) \chi(t - \tau) d\tau. \tag{3.35}$$

[21]For a given signal, the power spectrum provides the portion of a signal's power (energy per unit time) as a function of frequency bins.

Now the FDT relates the *power spectrum*[21] to the imaginary part of the *Fourier transform* (see Chapter 4 for an introduction to complex numbers) of the susceptibility $\chi(t)$, i.e. $\hat{\chi}(\omega) = \mathrm{Re}\hat{\chi}(\omega) + i\mathrm{Im}\hat{\chi}(\omega)$,

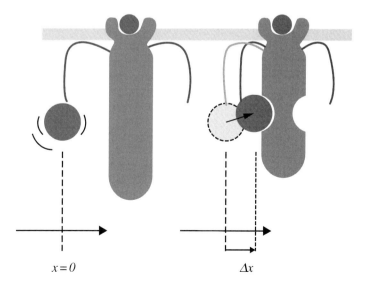

**Fig. 3.8** Receptor with tethered protein at position $x$. The protein (large circle) can either be unbound (left) or bound (right) to the main part of receptor (large object). The bound state may be induced by ligand binding (small circle) to the extracellular domain of the receptor.

$x = 0$

$\Delta x$

$$S(\omega) = \frac{2k_B T}{\omega}\mathrm{Im}\hat{\chi}(\omega),\qquad(3.36)$$

where the power spectrum $S(\omega)$ describes the frequency content of a time-varying process and helps to identify periodicities. The formal derivation of the FDT is provided in the Appendix A.

## Chapter summary

- The root-mean square displacement of diffusion in $n$ dimensions is given by rms $= (2nDt)^{1/2}$. Hence, diffusion is effective over short distances but not over long distances.
- The diffusion equation $\frac{\partial c(x,t)}{\partial t} = D\frac{\partial^2 c(x,t)}{\partial x^2}$ can be derived from Fick's first and second laws.
- The Boltzmann distribution is given by $p(E_i) = \frac{g_i e^{-E_i/k_B T}}{Z}$ with $g_i$ the degeneracy of levels with energy $E_i$ and $Z$ the partition function. This probability distribution describes how particles are distributed over energy levels at thermal energy $k_B T$.
- The fluctuation–dissipation theorem $S(\omega) = \frac{2k_B T}{\omega}\mathrm{Im}\hat{\chi}(\omega)$ relates equilibrium fluctuations to the non-equilibrium linear response.

## Further reading

(1) *Physical Biology of the Cell*, by Rob Phillips, Jane Kondev, and Julie Theriot (Garland Science, 1st Edition, 2009).

(2) *Random Walks in Biology*, by Howard C. Berg (Princeton, 1993).

(3) *Soft Condensed Matter*, by Richard A. L. Jones (Oxford University Press, 2002).

(4) *Biological Thermodynamics*, by Donald T. Haynie (Cambridge University Press, 2001).

(5) *Statistical Thermodynamics of Nonequilibrium Processes*, by Joel Keizer (Springer Press, 1987).

(6) *Random Walks and Random Environments. Volume 1: Random Walks*, by Barry D. Hughes (Clarendon Press, Oxford, 1995).

# 4 Mathematical tools

In this chapter we cover the mathematical tools that we use in the subsequent chapters of this book. The list of tools presented here and in Appendix A is fairly representative, occurring in a large number of biological problems, even outside the scope of this book. Someone familiar with mathematical modeling may skip this chapter during a first read.

Why do we need mathematics in biology? Mathematics is necessary in order to make biology quantitative—that is, precise and predictive. An important goal of mathematical modeling is to accurately describe and predict biological processes such as the concentrations of molecules in a cell or in a biochemical pathway in time. However, a couple of decisions need to be made by the modeler before doing so. Unlike hard-wired microprocessors in computers, biological processes inside a living cell are noisy (random). Hence, to mathematically describe the time evolution of the copy numbers of molecular species, two different approaches are generally possible. First, there are deterministic approaches, such as differential equations, which predict the average concentrations of molecules in time. Second, there are stochastic approaches, which include the fluctuations of molecular species in time, often requiring simulations instead of precise analytical or numerical solutions.

In addition to dynamical rate equations, we can use completely different mathematical approaches to describe biological pathways. Information theory treats biological processes from the perspective of information transmission familiar from communications research. This useful tool will be introduced at the end of the chapter.

## 4.1 Ordinary differential equations

When the copy number of a molecule in a cell is large, e.g. if we deal with an important signaling molecule or an abundant protein like ribosomes, we can approximate the discrete copy number of a molecular species by a continuous variable, denoted e.g. by $n$. Its rate of change is most simply described by an ordinary differential equation of the form

$$\frac{dn}{dt} = \underbrace{k}_{\text{source}} - \underbrace{\gamma \cdot n}_{\text{decay}}, \qquad (4.1)$$

where the positive first term on the right-hand side describes the source of the molecule (a constant rate of production $k$), and the negative second term on the right-hand side describes the degradation, decay, or dilution of the molecule. Here, $\gamma \cdot n$ means that each molecule has a constant rate of degradation (inverse lifetime $\gamma$), so the total rate of degradation is the larger the more molecules there are. In other words, this term is equivalent to radioactive decay, well known from physics. Note that while Eq. (4.1) is about the copy number of molecules, we could also divide Eq. (4.1) by the reaction volume $V$ and write the equation in terms of concentration $n$ (note we use the same symbol for the continuous copy number and concentration), that is $dn/dt = k/V - \gamma \cdot n = \tilde{k} - \gamma n$ with $\tilde{k} = k/V$.

To solve the example of Eq. (4.1) analytically, we introduce the method of *separation of variables*. We start with writing Eq. (4.1) differently

$$\frac{dn}{dt} = \frac{\bar{n} - n}{\tau} \tag{4.2}$$

with the old parameters given by $k = \bar{n}/\tau$ and $\gamma = \tau^{-1}$. Now, we separate the variables $n$ and $t$ to opposite sides of the equation via

$$\frac{dn}{\bar{n} - n} = \frac{dt}{\tau}, \tag{4.3}$$

which can now be integrated to yield

$$n(t) = \bar{n}(1 - e^{-t/\tau}). \tag{4.4}$$

For small times $t$, we can expand to first order in $t/\tau$, i.e. $e^{-t/\tau} \approx 1 - t/\tau$, and obtain an approximate linear relationship in time $n \approx \bar{n}[1 - (1 - t/\tau)] = \frac{\bar{n}}{\tau} \cdot t = kt$. For large $t$, Eq. (4.4) produces the steady-state concentration $\bar{n} = k/\gamma$.

Alternatively, we use *Euler's method* to numerically solve Eq. (4.1), now written as

$$\frac{dn}{dt} = f(n), \tag{4.5}$$

where the right-hand side of Eq. (4.1) is given in terms of the $n$-dependent function $f(n)$. Using initial condition $n = n_0$ at $t = t_0$, we can approximate the solution in a systematic way using a vector-field interpretation: at $n_0$, quantity $f(n_0)$ is like a velocity of a fictitious particle at position $n_0$ in one dimension. In time interval $\Delta t$, such a particle moves by $f(n_0)\Delta t$ (distance=rate×time).[1] As a consequence, the new position of the particle is

$$n(t_0 + \Delta t) = n_1 \approx n_0 + f(n_0)\Delta t. \tag{4.6}$$

Now we iterate, using $n_1$ as the old position and $f(n_1)$ as the velocity. This leads to the new position

$$n_2 = n_1 + f(n_1)\Delta t, \tag{4.7}$$

and more generally to

$$n_{i+1} = n_i + f(n_i)\Delta t, \tag{4.8}$$

[1] For this analogy, the velocity corresponding to position $x$ is $dx/dt = v$. If $v$ were constant—that is, independent of time—then $x = x_0 + vt$ with $x_0$ the initial position. In our example in the text the "velocity" depends on $x$, so we need to integrate (or iterate) instead.

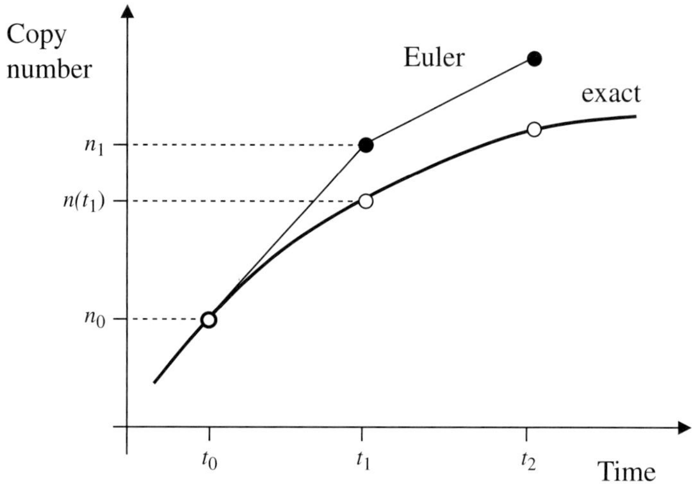

**Fig. 4.1** Illustration of numerical Euler's method to solve an ordinary differential equation. Shown are the Euler approximation for a molecule copy number, calculated iteratively on discrete time points, as well as the exact curve for comparison. The approximation becomes more and more inaccurate with time.

where subscript $i$ counts the iteration cycle.

As illustrated in Fig. 4.1, Euler's method has one severe disadvantage. To avoid large errors, the time step $\Delta t$ needs to be very small. However, if the time step is very small, one requires many iterations, and consequently, the solution progresses only very slowly and suffers from round-off errors. Additionally, the derivative $dn/dt$ is only calculated at the left-hand side of time interval $\Delta t$. To remedy these shortcomings, the *improved Euler's method*, or the even better, the *Runge—Kutta method* can be used.

## 4.2  Kinetic laws

### 4.2.1  Law of mass action

The "law" states that the rate of an elementary reaction (a reaction that proceeds through only one transition state, i.e. one mechanistic step) is simply proportional to the product of the concentrations of the participating molecules. For instance, the reaction between an enzyme of concentration $E$ and a substrate of concentration $S$, which produces a product of concentration $P$, can principally be reversible and described by

$$E + S \underset{k_{-1}}{\overset{k_1}{\rightleftharpoons}} E + P \tag{4.9}$$

(at least for an inefficient enzyme). By definition, the enzyme is not modified in this reaction. In this reaction, the forward and backward rates are given by the products of the concentrations of the reacting species, multiplied by the rate constant

$$r_+ = k_1 \cdot E \cdot S \tag{4.10}$$

$$r_- = k_{-1} \cdot E \cdot P. \tag{4.11}$$

Using these rates, the rate of change of the product can be described by the ordinary differential equation

$$\frac{dP}{dt} = k_1 E \cdot S - k_{-1} E \cdot P. \tag{4.12}$$

## 4.2.2   Michaelis–Menten kinetics

This type of kinetics can describe a large number of enzymes reasonably well (more realistically than Eq. (4.9)), and is relevant for situations where very simple kinetics can be assumed (without intermediate or product inhibition, and without allostery or cooperativity). More complex models exist for the cases where the assumptions of Michaelis–Menten kinetics are no longer appropriate. The reaction is given by

$$E + S \underset{k_{-1}}{\overset{k_1}{\rightleftharpoons}} ES \xrightarrow{k_2} E + P, \tag{4.13}$$

with $ES$ an intermediate complex made of $E$ and $S$.

The validity of the following derivation rests on two key assumptions: (1) the total enzyme concentration and the concentration of the intermediate complex do not change over time. (2) The enzymatic reaction is assumed to be irreversible.

In this model, what does the important rate of change of the product, $dP/dt$, look like? While the product is the measurable quantity, we would like to remove the intermediate $ES$ from the formula $dP/dt = k_2 \cdot ES$ as it is normally not measured. Making the quasi-steady-state assumption that the concentration of the intermediate complex $ES$ changes much more slowly than the concentrations of the product and substrate, we obtain

$$\frac{dES}{dt} = 0 = k_1 E \cdot S - (k_1 + k_2)ES \tag{4.14}$$

$$\Longrightarrow ES = \frac{k_1 E \cdot S}{k_{-1} + k_2} = \frac{E \cdot S}{K_M}, \tag{4.15}$$

where $K_M = (k_{-1} + k_2)/k_1$ is the so-called Michaelis–Menten constant. For very small $k_2$, $K_M \approx K_D = k_{-1}/k_1$ is equilibrium dissociation constant.

Using the total concentration $E_0 = E + ES$ of the enzyme, where $E$ is the free and $ES$ is the bound enzyme, we obtain for the bound enzyme concentration

$$ES = \frac{(E_0 - ES)S}{K_M} \tag{4.16}$$

$$\Longrightarrow ES = E_0 \frac{S}{S + K_M}. \tag{4.17}$$

Using this expression for $ES$, we are finally able to calculate the rate of change of product $P$, often also called the velocity $v$ of the reaction, via

$$v = \frac{dP}{dt} = k_2 \cdot ES = k_2 E_0 \frac{S}{S + K_M} = v_{\max} \frac{S}{S + K_M} \tag{4.18}$$

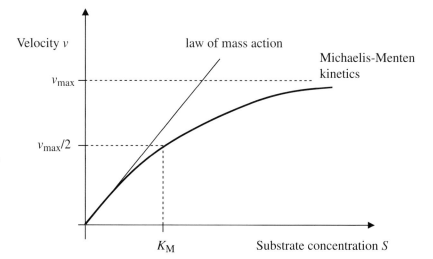

**Fig. 4.2** Michaelis–Menten kinetics. The plot illustrates the dependence of the reaction velocity on the substrate concentration. At large substrate concentration the reaction velocity saturates. The Michaelis–Menten kinetics are compared with the law of mass action, which corresponds to small substrate concentrations.

with $v_{\mathrm{max}} = k_2 E_0$ the maximal velocity of the reaction. For little available substrate, i.e. small $S$, we approximately obtain mass-action-type kinetics

$$v \approx \frac{v_{\mathrm{max}}}{K_M} \cdot S = \frac{k_2 E_0}{K_M} \cdot S. \tag{4.19}$$

Fig. 4.2 shows the reaction velocity as a function of substrate concentration $S$, including the asymptotic behavior for small and very large concentrations of the substrate. For a large concentration $S$, the enzyme saturates and the velocity reaches a maximum. Note, however, that $S$ generally changes during the reaction, so often people just calculate the initial velocity for a given $S$.

## 4.3   Master equation

If the number $n$ of a molecule is large, we can ignore the fact that $n$ is an integer in the dynamical rate equation

$$\frac{dn}{dt} = k - \gamma \cdot n = f_n - g_n, \tag{4.20}$$

where we called the source term $f_n$ and the decay term $g_n$. However, if the number is small, e.g. if a gene is transcribed into a few copies of mRNA molecule, we cannot ignore the discreteness of $n$, and a more general approach based on the chemical master equation is needed. The master equation describes the reaction kinetics probabilistically, such as for an ensemble of cells.

To derive the master equation, note that in our example molecule numbers can only change by one at a time. While the number of reactions occurring with some rate $r$ (such as $k$) in time $\tau$ is $r\tau$ and hence macroscopically large, we can divide time $\tau$ into $M$ small bins, leading to $rT/M$ reactions per bin. Thus with $M$ large enough, no more than one reaction will occur per bin, demonstrating the applicability of a simple one-step process.

Using variable $N$ for integer copy numbers, we can now write the following master equation:

$$\frac{dp_N}{dt} = -(f_N + g_N)p_N + f_{N-1}p_{N-1} + g_{N+1}p_{N+1}, \qquad (4.21)$$

where $p_N$ is the probability to have exactly $N$ molecules. Factors $f_N$, $f_{N-1}$, $g_N$, and $g_{N+1}$ are called transition probabilities. Importantly, the first, negative term in Eq. (4.21) describes the decay of the state with $N$ molecules, either by adding one molecule with rate $f_N$ or by removing one with rate $g_N$. The second and third positive terms describe the production of one molecule, either by creating a molecule from state $N - 1$ (term 2) or by destroying one from state $N + 1$ (term 3). Eq. (4.21) is illustrated in Fig. 4.3. While Eq. (4.21) is linear in $p_N$ and hence simple, it constitutes an infinite set of coupled differential equations, making it impossible to solve exactly in most cases.[2]

### 4.3.1   Emergence of the deterministic law

The deterministic approach, Eq. (4.1), describes the average number of particles, while the master equation, Eq. (4.21), describes the probability distribution, characterized by the average (first moment), the variance (second moment), and higher moments. Fortunately, these can generally be obtained without solving the master equation explicitly. In the following, we derive the first moment and rediscover the deterministic law.

Using notation $\langle N \rangle = \sum_{N=0}^{\infty} N p_N$ for the average number of molecules, we obtain from the master equation

$$\frac{d\langle N \rangle}{dt} = -k \sum_{N=0} N p_N - \gamma \sum_{N=0} N^2 p_N + k \sum_{N=0} N p_{N-1} + \gamma \sum_{N=0} N(N+1)p_{N+1}. \qquad (4.22)$$

The third and forth terms on the right-hand side of Eq. (4.22) can be expressed in terms of $p_N$, allowing us to finally obtain

$$\frac{d\langle N \rangle}{dt} = k - \gamma \langle N \rangle. \qquad (4.23)$$

At steady-state, the time derivative is zero, and we obtain[3]

$$\langle N \rangle_s = k/\gamma. \qquad (4.24)$$

[2] If transition probabilities are constant or linear in $n$ (like in our example), moment generating functions can be used to solve the master equation. These functions turn an infinite set of ordinary differential equations into a single partial differential equation.

[3] Nomenclature: note that $\langle N \rangle$ in Eq. (4.23) and $\langle N \rangle_s$ in Eq. (4.24) correspond to respective $n$ and $\bar{n}$ in Eq. (4.2). The new notation with $\langle \dots \rangle$ is more appropriate for stochastic processes.

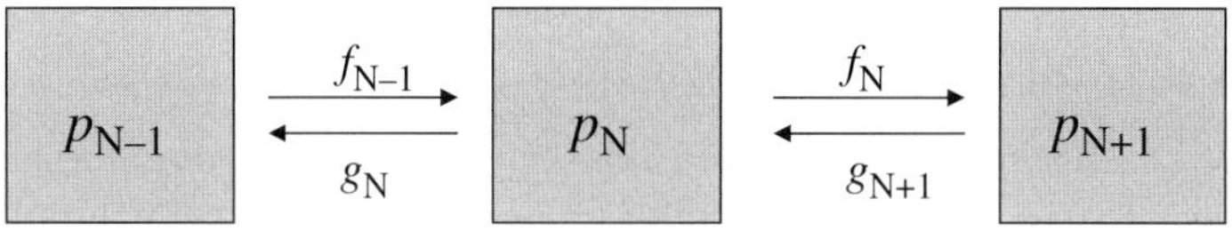

Fig. 4.3 Illustration of the one-step master equation. Processes affecting $p_N$, the probability to observing $N$ molecules, only depends on $p_{N-1}$ and $p_{N+1}$. Rates $f$ and $g$ correspond to production and degradation, respectively.

## 4.4   Poisson distribution

The Poisson distribution is a discrete probability distribution that describes the probability that a number of events occurs in a fixed period of time, assuming events occur with a known average rate and independently of previous events. Since these assumptions are often fulfilled in biological systems, the distribution is very important for us. Examples include the basal expression of a protein independent of other regulator proteins or the spontaneous occurance of mutations in a DNA molecule. The Poisson distribution can also be used for the number of events in other specified intervals such as distance, area, or volume. An example might be the locations of a freely diffusing molecule in a cell at any point in time.

To illustrate the Poisson distribution, we derive the steady-state probability distribution of Eq. (4.22) and show that it is Poissonian. At steady state, all time derivatives are zero, and we have

$$0 = -(f_N + g_N)p_N + f_{N-1}p_{N-1} + g_{N+1}p_{N+1}. \tag{4.25}$$

Dividing by $\gamma$ and using short notation $\bar{n}$ for the steady-state value, we obtain

$$-(\bar{n} + N)p_N + \bar{n}p_{N-1} + (N+1)p_{N+1} = 0 \tag{4.26}$$
$$\implies \quad (N+1)p_{N+1} - \bar{n}p_N = Np_N - \bar{n}p_{N-1} = 0. \tag{4.27}$$

The last step comes from the fact that the relation is valid for all $N$, including $N = 0$ (note that $p_{N-1} = 0$). Rearranging terms, we obtain the recursive formula

$$p_N = \frac{\bar{n}}{N}p_{N-1} = \frac{\bar{n}}{N} \cdot \frac{\bar{n}}{(N-1)}p_{N-2} = \cdots = \frac{\bar{n}^N}{N!}p_0. \tag{4.28}$$

As $p_0$ is not specified, we need to obtain its value from normalization. Setting the sum over all $p_N$ equal to one, we obtain

$$\sum_{N=0} p_N = p_0 \sum_{N=0} \frac{\bar{n}^N}{N!} = p_0 e^{\bar{n}} = 1 \ \rightarrow \ p_0 = e^{-\bar{n}}, \tag{4.29}$$

and for the final probability distribution

$$p_N = \frac{\bar{n}^N}{N!}e^{-\bar{n}}. \tag{4.30}$$

[4]While the Poisson distribution is only characterized by the average molecule number, the Gaussian distribution depends both on the average and variance. For sufficiently large averages, the Gaussian distribution (with the same average and variance) is an excellent approximation to the Poisson distribution.

This Poissonian result is expected as the rates of production and degradation (per molecule) are constant, not depending on the number of molecules, nor their history. Importantly, for a Poisson distribution, the variance $\langle(\delta N)^2\rangle = \langle(N - \langle N\rangle)^2\rangle = \langle N^2\rangle - \langle N\rangle^2$ is equal to the mean $\langle N\rangle$, i.e. $\langle(\delta N)^2\rangle = \langle N\rangle$.[4] This will prove usefull next.

### 4.4.1   Limit of large numbers

We said earlier that when noise is small, ordinary differential equations can be used to describe the kinetics of the average copy number of molecules. Using the Poisson distribution, we can be more specific about what we mean by small noise or fluctuations in molecule copy numbers. Since for a Poisson distribution $\langle (\delta N)^2 \rangle = \langle N \rangle$, and hence the absolute size of the fluctuations increases with the average, the relative fluctuations

$$\frac{\sqrt{\langle (\delta N)^2 \rangle}}{\langle N \rangle} = \frac{1}{\sqrt{\langle N \rangle}} \tag{4.31}$$

decrease with increasing average. Consequently, for very large numbers of molecules, the relative fluctuations can be neglected, and the kinetics of the average is an excellent approximation for the time evolution of the system. As an example, if the average is 20 molecules, then the relative fluctuations are 22%. If the average is 500, then the relative fluctuations are only 4%.

## 4.5   Waiting-time distribution

What about the time intervals between successive occurrences of a reaction, and how are they distributed? Suppose a reaction occurs with rate $r$. Then the probability that the reaction occurs in time interval $\delta t$ is $P(\text{event in } \delta t) = r\delta t$, and that it does not occur is $P(\text{no event in } \delta t) = 1 - r\delta t$. This can be used to calculate the distribution that the event occurs in time interval $[t, t + \delta t]$, but not earlier. For this purpose, we imagine the time to be divided into $M$ small bins of size $\delta t$ as shown in Fig. 4.4. In this case we can apply the probabilities already given, and obtain[5]

[5]Note $\lim_{N \to \infty} (1 + x/N)^N = e^x$.

$$P(t) = P(\text{event in } [t, t + \delta t]) = \underbrace{(1 - r\delta t) \dots (1 - r\delta t)}_{M \text{ times}} \cdot r\delta t$$

$$= \left(1 - r\frac{t}{M}\right)^M \cdot r\delta t \approx e^{-rt} r\delta t. \tag{4.32}$$

As a result, waiting times between successive events are exponentially distributed. In fact, knowing $P(t)$ provides a simple numerical procedure to simulate stochastic systems, called *Gillespie simulation* (see Appendix A). Specifically, $P(t)$ tells us when reactions occur given the total rate of all possible reactions, and once a reaction occurs, how to update the simulation time.

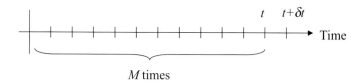

*M* times

**Fig. 4.4** To derive the waiting-time distribution of a chemical reaction, we divide the time into small time intervals of length $\delta t$ ("bins") so there is at most one occurance of the reaction in any bin.

## 4.6   Langevin small-noise approximation

When the number of molecules is large and the fluctuations of the molecule numbers around their average are small, we can use a very useful approximation of including the noise in the mathematical rate equations. Here we discuss in more detail the Langevin approach, already mentioned in Chapter 3 as a way to describe diffusion. However, the approach is general and very useful for a large number of biological problems.

Let us return to our simple model given by

$$\frac{dn}{dt} = k - \gamma n + \eta_n, \tag{4.33}$$

where $n$ is the continuous copy number of a molecule, created with constant rate $k$ and degraded (or diluted) with rate constant $\gamma$. In addition to Eq. (4.1), we now also included a noise term $\eta_n(t)$ to describe fluctuations around the average rate value. Hence, stochastic differential equations such as Eq. (4.33) are a middle way between ordinary-differential equations without any noise and the full stochastic description by the master equation. An additional advantage is that they can easily be solved numerically with, e.g., the Euler method. In the following we first describe how to estimate the noise term $\eta_n$.

The noise term $\eta_n$ is really the sum of two terms, $\eta_1$ and $\eta_2$, which are respectively associated with rates $k$ and $\gamma n$. We assume $\eta_1$ and $\eta_2$ to be independent on average, i.e. $\langle \eta_1(t)\eta_2(t)\rangle = 0$. In general, this is justified as different reactions are catalyzed by different proteins. Furthermore, we assume they are represented by so-called Gaussian white-noise terms with zero average, $\langle \eta_1 \rangle = \langle \eta_2 \rangle = 0$, and auto-correlation function

$$\langle \eta_j(t)\eta_j(t')\rangle = Q_j \cdot \delta t^{-1} \cdot e^{-|t-t'|/\delta t} \tag{4.34}$$
$$\rightarrow Q_j \cdot \delta(t-t') \tag{4.35}$$

where $Q_j$ the noise intensity and the limit in Eq. (4.35) is obtained for $\delta t \rightarrow 0$ and $Q_j \delta t = const$. Note that this noise is "Gaussian" distributed, since it is solely defined by the mean and variance, or more generally the auto-correlation function. It is "white" since the $\delta$-function in the time domain corresponds to an infinitely broad (flat) function in the frequency domain.

The noise intensities can be calculated assuming Poisson processes with average rates $k$ and $\gamma n$, resulting in $Q_1 = k$ and $Q_2 = \gamma n$. The former expression is the average number of molecules created per unit time, while the latter is the average number of molecules destroyed per unit time. The noise intensity of the total noise $\eta_n(t)$ is the sum, $Q = Q_1 + Q_2$, because we assume the two types of noise are independent.[6] In Appendix A we will derive the noise strengths from a systematic expansion of the master equation.

[6] As this point we notice that the variance can only equal the mean if we work with copy numbers and not with concentrations!

# 4.7 Information theory

The problems we encounter in telecommunications and electronics are not very different from the problems that a biological cell faces in signal transduction. In both cases, stimuli or messages need to be encoded and transmitted over at times large distances, protected from noise corruption, and decoded at the receiver end. Once decoded, an appropriate response is normally launched. Information theory is very useful in this respect.

Suppose we have a random variable $x$, produced by a random but nevertheless well-defined process. We ask how much information is obtained when receiving a particular value. The amount of information can be regarded as the 'degree of surprise' on learning the outcome. For the outcome of an improbable event we are very surprised, and hence have learned a lot. In contrast, for a very probable result we are little surprised and have learned little.

A general measure of information content $h(x)$ should therefore depend on the probability of observing $x$. Specifically, it should depend monotonically on the probability distribution $p(x)$. Furthermore, the information of observing two independent events $x$ and $y$ should add up $h(x, y) = h(x) + h(y)$. Since for such events $p(x, y) = p(x)p(y)$, the logarithm has the desired properties. Specifically, we use

$$h(x) = -\log_2 p(x), \tag{4.36}$$

where the negative sign ensures that that information is positive. The choice of the base 2 of the logarithm is arbitrary but provides the information in units of bits. Information of one bit means a choice of two possible outcomes for a random variable.

To calculate the average amount of information over several trials of transmitting information, we average $h(x)$ over the probability distribution $p(x)$ and obtain Shannon's entropy formula[7]

$$H[x] = -\sum_x p(x) \log_2 p(x). \tag{4.37}$$

[7] A note on terminology. In Eq. (4.37) brackets [$x$] on the left-hand side mean that $x$ is a random variable, while $x$ on the right-hand side denotes a specific value of that random variable.

To show that the definition of entropy makes sense, let us consider a couple of simple examples. First, for $p(x) = 0$, $p(x) \log_2 p(x) = 0$ and an impossible event logically contributes no information. Second, a Gaussian distributed random variable, $p(x) = 1/\sqrt{2\pi\sigma^2} \exp\{-(x - \mu)^2/(2\sigma^2)\}$ with average $\mu$ and variance $\sigma^2$, has the entropy $H = 1/2\{1 + \ln(2\pi\sigma^2)\}$. Hence, as expected, the broader the distribution (large $\sigma$) the larger the entropy. This concept can also be extended to continuous variables. In this case, sums are replaced by integrals, although the latter are technically diverging due to the infinitely many possible outcomes.

Shannon's entropy has many applications. Let us consider the important case of two sets of variables, $x$ and $y$, desribed most generally by the joint probability distribution $p(x, y)$. The *mutual information* between these two variables is given by

$$I[x, y] = -\sum_x \sum_y p(x, y) \log_2 \left( \frac{p(x)p(y)}{p(x, y)} \right) \qquad (4.38)$$

with $p(x, y) = p(y|x)p(x) = p(x|y)p(y)$ in terms of conditional probabilities. This equation describes the degree of dependence of the two variables, or how much we know about one when we learned about the other. We have $I[x, y] \geq 0$ with the equal sign only for $x$ and $y$ independent with $p(x, y) = p(x) \cdot p(y)$. The mutual information can also be rewritten in the form $I[x, y] = H[x] - H[x|y] = H[y] - H[y|x]$ with $H[x|y]$ the conditional entropy of $x$ given outcome $y$, etc. This clearly shows that the mutual information is the reduction in uncertainty about $x$ when we know about $y$, and vice versa.

The mutual information can be cast into different forms, depending on the application. The *channel capacity* places an upper bound on the amount of information that can be reliably transmitted over any physical system. This channel capacity, defined as the mutual information rate $C[x, y] = \lim_{T \to \infty} I[x, y]/T$, is given by

$$C[x, y] = \frac{1}{4\pi} \int_{-\infty}^{\infty} d\omega \ln \left[ 1 + \frac{S(\omega)}{N(\omega)} \right], \qquad (4.39)$$

where $S(\omega)/N(\omega)$ is the frequency-dependent *signal-to-noise ratio* (SNR). Importantly, this result for the channel capacity shows that the mutual information and SNR are linked. Maximizing the mutual information is equivalent to maximizing the SNR. Note that Eq. (4.39) is technically only correct for so-called Gaussian channels with the signal and the noise fluctuating around their respective averages with Gaussian statistics.

The mutual information also leads to the *rate distortion theory*, which deals with lossy information compression. The rate distortion function $R(D)$ determines the minimal rate in bits that should be communicated over a channel, so that the input signal can be approximately reconstructed from the output signal without exceeding a given distortion (loss) $D$

$$R(D) = \min_{p(y|x) \text{with} \langle \tilde{D} \rangle \leq D} I[x, y]. \qquad (4.40)$$

That is, we minimize the mutual information over the conditional probability $p(y|x)$ of mapping inputs to (compressed channel) outputs, given that the expected distortion $\langle \tilde{D} \rangle$ must not be larger than $D$. For a Gaussian channel, we can provide an intuitive analytical expression. If we assume that $p(x)$ is Gaussian distributed with variance $\sigma_x^2$ and successive samples are statistically independent, then the rate distortion function is given by

$$R(D) = \begin{cases} \frac{1}{2} \log_2(\sigma_x^2/D) & \text{if } D \leq \sigma_x^2 \\ 0 & \text{if } D > \sigma_x^2 \end{cases} \qquad (4.41)$$

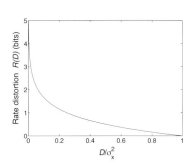

**Fig. 4.5** Rate distortion function $R(D)$ as a function of $D/\sigma_x^2$, the ratio of allowed distortion and Gaussian input noise with variance $\sigma_x^2$.

This rate distortion function is plotted in Fig. 4.5. The result says that the minimal rate of information that needs to be communicated goes up with smaller and smaller acceptable distortion $D$ (or increasing uncertainty $\sigma_x$), and that no compression exists that performs above the plotted curve.

# Chapter summary

- Modeling the dynamics of biological systems broadly includes deterministic and stochastic approaches. Deterministic approaches encompass ordinary and partial differential equations, stochastic approaches include stochastic differential equations (e.g. Langevin equations), chemical master equations, and their simulations by Gillespie algorithm and other kinetic Monte Carlo methods.

- Few differential equations can be solved analytically, but all can be integrated numerically. For most cases, the master equation cannot be solved analytically, but the time evolution of the moments or the steady-state probability distribution of the molecule numbers can sometimes be extracted.

- Information theory from communication research is another useful tool for analysing and interpreting biological pathways and decision-making. Specifically, the mutual information between two quantities describes how much we know of one quantity given the other one. Specific applications are the channel capacity and rate distortion theory.

# Further reading

(1) *Nonlinear Dynamics and Chaos*, by Steven H. Strogatz (Westview Press, 1994).

(2) *Stochastic Processes in Physics and Chemistry*, by Nico G. Van Kampen (Elsevier, 1981).

(3) *Pattern Recognition and Machine Learning*, by Christopher M. Bishop (Springer, 2006).

(4) *Measurements and their Uncertainties—A Practical Guide to Modern Error Analysis*, by Ifan G. Hughes, and Thomas P.A. Hase (Oxford, 2010)

(5) *Numerical Recipes*, by William H. Press, Saul A. Teukolsky, William T. Vetterling, and Brian P. Flannery (Cambridge University Press, 3rd Edition, 2007).

(6) *The Mathematical Theory of Communication*, by Claude E. Shannon (University of Illinois Press, 1949).

(7) *Elements of Information Theory*, by Thomas M. Cover and Joy A. Thomas (John Wiley and Sons, New York, 1991).

(8) 'An Information-theoretic characterization of the optimal gradient sensing response of cells', by Burton W. Andrews and Pablo A. Iglesias, *PLoS Comput Biol.* 3: e153 (2007).

(9) 'Mutual information between input and output trajectories of biochemical networks', by Filipe Tostevin and Pieter Rein ten Wolde, *Phys. Rev. Lett.* 102: 218101 (2009).

# Signal amplification and integration

## 5

The bacterium *E. coli* is able to sense minute changes of chemicals such as attractants in its environment. For instance, an increase in receptor occupancy by just 0.2% has been estimated to result in a 23% change in the bias of the motor between CW and CCW rotation. For such a noticeable effect on the downstream swimming behavior, weak external stimuli need to be amplified (in this example by a factor of 100). However, before tackling the amplification by chemoreceptor complexes and to gain some more intuition, we start with the well-known molecule hemoglobin. This example will allow us to introduce cooperative ligand binding by a multi-protein complex.

Hemoglobin is a very important molecule in our body, as it has the property of binding oxygen for its transport in the blood. The protein is evolutionarily designed in such a way that it can bind up to four oxygen molecules in switch-like (all-or-none) fashion. At low oxygen concentrations, no oxygen molecules bind, while at high oxygen concentrations, all four binding sites are occupied by oxygen molecules. The measured Hill coefficient, the previously introduced empirical measure of cooperativity, is about 3. What are the advantages of cooperativity, and where does cooperative behavior in molecules come from? Once these questions are clarified, we will first discuss a single receptor and subsequently receptor complexes.

## 5.1 Cooperativity by allostery

Hemoglobin is thought to have two conformational states as shown in Fig. 5.1, a tense (T) and a relaxed (R) state. While there is an energy barrier between the two states, due to the availability of thermal energy, such a molecule will generally switch between the two states randomly, establishing an equilibrium. However, this equilibrium can shift to one or the other state, increasing the stability and lifetime of the favored state. In hemoglobin the equilibrium is regulated by the oxygen concentration in the blood. Since only the R-state binds oxygen favorably, the R state is consequently favored at high oxygen concentration. Binding of one oxygen molecule induces a change in protein conformation, making it easier for other oxygen molecules to bind. This is the allosteric effect—a binding event on one side of the protein complex can affect the binding of another ligand molecule elsewhere on the complex. As a result, the

Relaxed (R)          Tense (T)

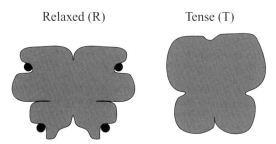

**Fig. 5.1** Cooperative binding of oxygen molecules (black dots) to hemoglobin. Both relaxed and tense protein conformations are shown.

average binding energy per oxygen molecule becomes larger and more favorable with each additional oxygen binding to the protein. Such cooperative binding has the advantage to effectively load and unload oxygen molecules from the blood to the cells in the tissues.

In addition to switch-like processes, there can also be graded responses, during which switching occurs more gradually and smoothly. As a rule of thumb, switch-like processes are often preferred by biological systems. A clear molecular state reflects a molecular decision, similarly to a bit of information in a computer's memory (0 or 1). Such a specific state can further be read out by other molecules, allowing the information to unambiguously be processed.

In 1965, Monod, Wyman, and Changeux proposed an allosteric model to describe cooperative oxygen binding to hemoglobin. This MWC model can be extended to address cooperative ligand binding to receptor complexes, able to describe the dose-response curves from FRET measurements, introduced in Chapter 2. The main assumptions are: (1) Receptors are two-state objects, which can either be in the *on* state, which signals (active), or in the *off* state, which does not signal to CheA (inactive). This two-state property is illustrated in Fig. 5.2. (2) A group of receptors is coupled to form a receptor complex. (3) Receptors are so strongly coupled that even the whole complex is a two-state object.

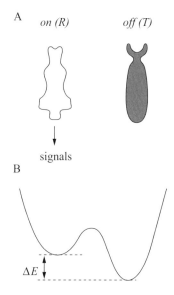

**Fig. 5.2** Illustration of a single two-state receptor. (A) Shown are the relaxed $R$ (or *on*) state on the left and the tense $T$ (or *off*) state on the right. (B) Energy double well diagram showing an energetic bias towards the off state by energy amount $\Delta E$.

## 5.2 Emergence of allostery from microscopic details

Fig. 5.3 shows the main kinetic features of the allosteric MWC model. Motivated by hemoglobin, let us consider a receptor complex of $N = 4$ receptors. On the left-hand side of this figure, the *on* (abbreviated by $R$) state is shown, for simplicity assumed not to bind ligand with significant strength. Hence, only the ligand-unbound state $R(0)$ is shown. On the right-hand side, the *off* (or $T$) state is shown, which can bind up to $N$ ligand molecules in consecutive order starting from the unbound state $T(0)$, including the state with one bound ligand $T(1)$, and so on up to state $T(N)$ with $N$ ligand molecules bound. In the ligand unbound state, the rate constant of switching from the *on* to the *off* state is $k_f$, while the reverse reaction has rate constant $k_b$. Once in the $T$ state, the complex can bind one ligand molecule with rate $Nk_+c$, where $k_+$ is the rate constant of binding and $c$ is the ligand concentration. The factor

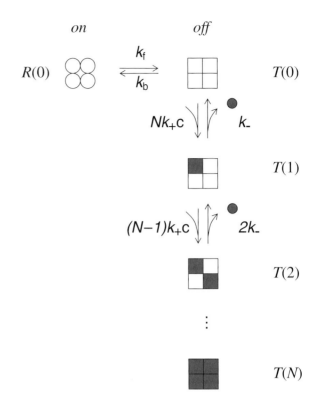

**Fig. 5.3** States of a receptor complex of four receptors. On the left side, the relaxed *on* (or $R$) state is shown, which is assumed to not bind ligand. On the right side, the tense *off* (or $T$) state is shown which can bind up to $N = 4$ ligand molecules. Also shown are switching rate constants, as well as ligand binding and unbinding rate constants as explained in the text.

$N$ arises since all receptors of the complex can bind ligand. Hence, for binding a second ligand molecule the rate is correspondingly $(N-1)k_+c$. Furthermore, the complex with one bound ligand can unbind this ligand molecule with rate $k_-$. In contrast, if two ligand molecules are bound, one ligand molecule can unbind with rate $2k_-$ as there are two possibilities for this unbinding to occur.

Next, we can write down the apparent ligand dissociation constants for the different $T$ states, the inverse of the ligand binding affinities, to see how the strength of binding changes along the sequence of states. Remember that for cooperativity to occur, the average ligand binding energy should become the larger the more ligand molecules are bound. The ligand dissociation constant of a single receptor in the $T$ state is $K_D^{\text{off}} = k_-/k_+$. In case of the complex, the apparent ligand dissociation constant for the first ligand molecule is $K_D^{\text{off}}(1) = k_-/(Nk_+) = K_D^{\text{off}}/N$ and hence is $N$ times smaller than for the single receptor. Curiously, from now on the ligand dissociation constant becomes larger (binding gets weaker) for more and more ligand molecules. For the second ligand molecule we obtain $K_D^{\text{off}}(2) = 2k_-/[(N-1)k_+] = 2K_D^{\text{off}}/(N-1)$, and finally for $N$th ligand molecule the constant is the largest—specifically, it is $N$ times as large as the single receptor constant $K_D^{\text{off}}(N) = Nk_-/k_+ = NK_D^{\text{off}}$. Hence, ligand binding starts out $N$ times stronger, equals the strength of a single receptor at half occupation, and becomes $N$ times weaker for the fully occupied complex. How can it be understood, then, that the average binding energy is larger, that is more negative, compared to a single receptor?

Due to the ratio $k_f/k_b$ of the forward and backward switching rate constants, there can be an energy bias $\Delta E$ between the $T$ and the $R$ state (see Fig. 5.2)

$$e^{N\Delta E} = e^{N(E_{\mathrm{on}} - E_{\mathrm{off}})} = \frac{k_f}{k_b} \tag{5.1}$$

with no bias ($\Delta E = 0$) corresponding to both rate constants being exactly equal. As a result, two regimes of operation are apparent. In regime I with $\Delta E > 0$ the complex is mostly in the $T$ state, while in regime II with $\Delta E < 0$ the complex is mainly in the $R$ state.

In Regime I the complex is expected to be highly sensitive to ligand binding. This is because the residual activity (probability to be in the $R$ state) is already small, and a single bound ligand molecule with apparent dissociation constant $K_D^{\mathrm{off}}(1) = K_D^{\mathrm{off}}/N$ is expected to switch the complex completely to the $T$ state. More specifically, we can exploit detailed balance: that is, at equilibrium each reaction in Fig. 5.3 is exactly balanced by its reverse reaction, so that

$$T(1) = \frac{Nk_+c}{k_-}T(0) = \frac{Nk_+c}{k_-}e^{N\Delta E}R(0) \tag{5.2}$$

$$= e^{N\Delta E + k_B T \ln\left(\frac{c}{K_D^{\mathrm{off}}/N}\right)}R(0), \tag{5.3}$$

where $c$ is the ligand concentration and $k_B T$ is the thermal energy. Both terms in the exponential have the same sign for $c > K_D^{\mathrm{off}}/N$ and contribute towards the $T$ state. However, this sensitive Regime I is not the cooperative Regime II known from hemoglobin.

In Regime II there is a strong bias towards the non-binding $R$ state in absence of ligand. Due to this strong bias the binding of a single ligand molecule will have little effect in stabilizing the $T$ state. If we consider the $N$-ligand bound complex, we have

$$T(N) = \left(\frac{c}{K_D^{\mathrm{off}}}\right)^N e^{N\Delta E}R(0) \tag{5.4}$$

$$= e^{N\Delta E + Nk_B T \ln\left(\frac{c}{K_D^{\mathrm{off}}}\right)}R(0), \tag{5.5}$$

where both terms in exponential have now opposite signs for $c > K_D^{\mathrm{off}}$ and hence work against each other. As a result, the average ligand binding energy is given by

$$E_N^{\mathrm{bind}} = |\Delta E| - k_B T \ln\left(\frac{c}{K_D^{\mathrm{off}}}\right), \tag{5.6}$$

which, due to the first term, is larger than the binding energy of a single ligand given by $E_1^{\mathrm{bind}} = N|\Delta E| - k_B T[\ln N + \ln(c/K_D^{\mathrm{off}})] \approx N|\Delta E| - k_B T \ln(c/K_D^{\mathrm{off}})$ using Eq. (5.3) and assuming $\ln 4 \approx 1$ is small compared to other terms. Hence, $N|\Delta E|$ acts like a penalty which needs to first be overcome to reach the $T$ state for binding ligand. This penalty is shared between all bound ligand molecules, so it is more favorable to bind as many ligand molecules as possible, leading to cooperative binding. Next, we will formally derive the dose-response curves to compare with FRET experiments.

## 5.3   Two-state equilibrium receptor model

[1]Based on Ref. (6).

Following the arguments from above, let us now apply rigorous statistical mechanics to this problem.[1] First we consider a single two-state receptor. From there we will move to a receptor complex.

As shown in Fig. 5.4, a single two-state receptor can be in four different microstates, assuming for now that binding can occur in both the *on* and *off* states: (1) *On* without a ligand molecule bound, (2) *on* with a ligand molecule bound, (3) *off* without a ligand molecule bound, and (4) *off* with a ligand molecule bound. The energies of these four states are given by

$$E_{\mathrm{on}} \tag{5.7}$$

$$E_{\mathrm{on}}^{L} = E_{\mathrm{on}} - k_B T \ln\left(\frac{c}{K_{\mathrm{D}}^{\mathrm{on}}}\right) \tag{5.8}$$

$$E_{\mathrm{off}} \tag{5.9}$$

$$E_{\mathrm{off}}^{L} = E_{\mathrm{off}} - k_B T \ln\left(\frac{c}{K_{\mathrm{D}}^{\mathrm{off}}}\right) \tag{5.10}$$

with $K_{\mathrm{D}}^{\mathrm{on}}$ and $K_{\mathrm{D}}^{\mathrm{off}}$ the ligand-dissociation constants for the receptor in the *on* and *off* state, respectively. The ligand-bound energies emerge from the definition of the binding-free energy, *e.g.* $E_{\mathrm{on}}^{\mathrm{bind}} = E_{\mathrm{on}}^{L} - E_{\mathrm{on}}$ for the *on* state, and the experimentally known relation between the ligand-dissociation constant and the binding-free energy given by $K_{D}^{\mathrm{on}} = c \cdot e^{E_{\mathrm{bind}}/(k_B T)}$. The deeper origin of the last relation lies in the competition between the release of binding energy and loss of mixing entropy upon ligand-receptor binding. Note also that in Eq. (5.8) the ligand-bound energy is negative (and hence favorable) for $c > K_{D}^{\mathrm{on}}$, equal to ligand-unbound energy $E_{\mathrm{on}}$ for $c = K_{D}^{\mathrm{on}}$, and positive (and hence unfavorable) for $c < K_{D}^{\mathrm{on}}$. (Analogous arguments apply for Eq. (5.10).)

To reconnect with the two-state property, the four energies can be combined into two: the energy of the *on* state with or without ligand, and the energy of the *off* state with or without ligand:

$$f_{\mathrm{on}} = E_{\mathrm{on}} - k_B T \ln\left(1 + \frac{c}{K_{\mathrm{D}}^{\mathrm{on}}}\right) \tag{5.11}$$

$$f_{\mathrm{off}} = E_{\mathrm{off}} - k_B T \ln\left(1 + \frac{c}{K_{\mathrm{D}}^{\mathrm{off}}}\right). \tag{5.12}$$

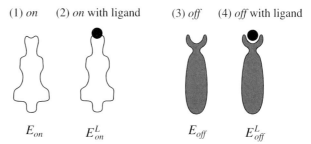

**Fig. 5.4** Energies and illustrative conformations of the four microstates of a receptor.

Note that technically we have been talking about free energies here, since the quantities introduced above depend on temperature and entropy. To see that the above expressions make sense, consider Eq. (5.11). For $c$ much smaller than $K_D^{\text{on}}$, the logarithm and hence ligand plays no role and what remains is the energy of the unbound state as expected. Next we demonstrate how the energy expressions Eqs. (5.11) and (5.12) are derived.

To calculate the probability that the receptor is *on*, we invoke Boltzmann statistics, which tell us that the probability of finding the system in a state with energy $E_i$ is $p_i = e^{-E_i/(k_B T)}/Z$, where $Z$ is the partition function defined by $Z = \sum_i e^{-E_i/(k_B T)}$ (see Chapter 3). In our example of the single receptor, the probability for the receptor to be *on* is given by

$$p_{\text{on}} = \frac{e^{-\frac{E_{\text{on}}}{k_B T}} + e^{-\frac{E_{\text{on}}^L}{k_B T}}}{e^{-\frac{E_{\text{on}}}{k_B T}} + e^{-\frac{E_{\text{on}}^L}{k_B T}} + e^{-\frac{E_{\text{off}}}{k_B T}} + e^{-\frac{E_{\text{off}}^L}{K_B T}}} \tag{5.13}$$

$$= \frac{e^{\frac{f_{\text{on}}}{k_B T}}}{e^{\frac{f_{\text{on}}}{k_B T}} + e^{\frac{f_{\text{off}}}{k_B T}}} \tag{5.14}$$

$$= \frac{1}{1 + e^{\frac{\Delta f}{k_B T}}}. \tag{5.15}$$

with the energies $f_{\text{on}}$ and $f_{\text{off}}$ given by Eqs. (5.11) and (5.12), respectively, and the energy difference between the *on* and *off* state provided by

$$\Delta f = \Delta E + k_B T \ln \left( \frac{1 + c/K_D^{\text{off}}}{1 + c/K_D^{\text{on}}} \right) \tag{5.16}$$

where $\Delta E = E_{\text{on}} - E_{\text{off}}$ is the energy difference in absence of ligand.

Evidently, the activity can be described by a single parameter $\Delta f = f_{\text{on}} - f_{\text{off}}$ as required for a two-state object. This energy difference depends on ligand-receptor binding as well as the receptor methylation level (via $\Delta E$). Since attractant inhibits the activity, ligand binding is taken to be stronger in the *off* state than in the *on* state of the receptor. Furthermore, receptor methylation favors the *on* state, as methylation increases the activity. This behavior is illustrated in Fig. 5.5, showing the energy profile of such a two-state receptor. Upon addition of attractant, the *off* state dominates with the energy difference increased ($\Delta f' > \Delta f$). The energy difference is reduced back to the original adapted value by receptor methylation.

Eq. (5.15) can be further simplified assuming that binding is very unfavorable in the *on* state, i.e. $K_D^{\text{on}}$ is very large. In this case we can write

$$p_{\text{on}} \approx \frac{1}{1 + e^{\frac{\Delta E}{k_B T}} \left( 1 + \frac{c}{K_D^{\text{off}}} \right)}. \tag{5.17}$$

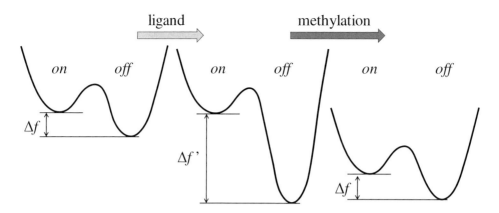

**Fig. 5.5** Illustration of the effects of ligand (attractant) binding and subsequent methylation on the receptor-complex energy profile.

## 5.4   Monod–Wyman–Changeux model for receptor signaling

Let us move from a single receptor to a receptor complex. If there are $N$ receptors in a complex, which are so strongly coupled, so they are always all *on* or *off* together, then Eq. (5.13) becomes

$$p_{\mathrm{on}} = \frac{\left(e^{-\frac{E_{\mathrm{on}}}{k_B T}} + e^{-\frac{E_{\mathrm{on}}^L}{k_B T}}\right)^N}{\left(e^{-\frac{E_{\mathrm{on}}}{k_B T}} + e^{-\frac{E_{\mathrm{on}}^L}{k_B T}}\right)^N + \left(e^{-\frac{E_{\mathrm{off}}}{k_B T}} + e^{-\frac{E_{\mathrm{off}}^L}{k_B T}}\right)^N} \tag{5.18}$$

$$= \frac{1}{1 + e^{\frac{\Delta F}{k_B T}}} \tag{5.19}$$

$$\approx \frac{1}{1 + e^{\frac{N\Delta E}{k_B T}}\left(1 + \frac{c}{K_D^{\mathrm{off}}}\right)^N} \tag{5.20}$$

where

$$\Delta F = N\left[\Delta E + k_B T \ln\left(\frac{1 + c/K_D^{\mathrm{off}}}{1 + c/K_D^{\mathrm{on}}}\right)\right] \tag{5.21}$$

is the energy difference of the whole MWC complex. For an homogeneous receptor complex, which contains only a single receptor type, this is simply $N$ times the energy difference $\Delta f$ of a single receptor. Note also that the $N$th power of the of *on* and *off* Boltzmann factors in Eq. (5.18) does not allow any mixed complexes in which some receptors are on and others are off in line with our strong receptor-receptor coupling assumption.

In Section 5.2 we mentioned two different regimes for a receptor complex depending on the energetic bias in absence of ligand. These two regimes translate into distinct signaling regimes, which are also evident from dose-response curves of receptors in specific modification states (see Fig. 2.7A). Unmodified receptors are highly sensitive and have a low activity at low ligand concentration (Regime I). In contrast, highly modified receptors are very insensitive, i.e. are very active at low attractant concentration, and exhibit large cooperativity (Regime II). In the following, we show that these two regimes are correctly predicted by the model.

In Regime I, receptors are largely inactive and hence $E_{\text{on}} \gg E_{\text{off}}$ (or $\Delta E \gg 0$). In this regime we can neglect the number 1 in the denominator of Eq. (5.20), and obtain

$$p_{\text{on}} \approx \frac{e^{-\frac{N\Delta E}{k_B T}}}{\left(1 + c/K_D^{\text{off}}\right)^N}, \tag{5.22}$$

leading to

$$p_{\text{on}}^{\max}(c = 0) = e^{-N\Delta E} \tag{5.23}$$

for the maximal activity at zero ligand concentration. Hence, clustered receptors in Regime I have an exponentially small activity. To derive the inhibition constant $K_i$, i.e. the ligand concentration for which receptor activity is half maximal, we set $p_{\text{on}} = e^{-\frac{N\Delta E}{k_B T}}/2$ in Eq. (5.22) and solve for $c$. This gives $K_i = K_D^{\text{off}} \ln 2/N$. Hence, a receptor complex is about $N$ times more sensitive than a single receptor. Note ligand binding to receptors in this regime is non-cooperative. According to Eq. (5.22),

$$p_{\text{on}} \approx e^{-\frac{N\Delta E}{k_B T}}\left(1 - Nc/K_D^{\text{off}}\right) \tag{5.24}$$

for $c \ll K_D^{\text{off}}$, which turns off for increasing $c$ without a threshold.

In Regime II, receptors are highly active and hence $E_{\text{on}} \ll E_{\text{off}}$ (or $\Delta E \ll 0$). Using $p_{\text{on}} = 1/2$ in Eq. (5.20) results in

$$e^{\frac{N\Delta E}{k_B T}}\left(1 + c/K_D^{\text{off}}\right)^N = 1 \tag{5.25}$$

and finally $K_i = K_D^{\text{off}} e^{\frac{|\Delta E|}{k_B T}}$, which increases with receptor methylation (or modification) since $|\Delta E|$ increases with methylation. Furthermore, for $c \gg K_D^{\text{off}}$, we obtain the following Hill-type equation

$$p_{\text{on}} \approx \frac{1}{1 + \left(e^{\frac{\Delta E}{k_B T}} \cdot \frac{c}{K_D^{\text{off}}}\right)^N}, \tag{5.26}$$

where the complex size $N$ plays the role of the Hill coefficient. Hence, receptors in Regime II are overall very insensitive to ligand with receptor methylation shifting the apparent ligand dissociation constant to larger and larger values. Hence in this regime clustering leads to a Hill-type equation, indicative of cooperative ligand binding and signaling.

The MWC model can very easily explain the features of dose-response curves. But what about other mechanisms to explain the data? Indeed, other models with, e.g., methylation-dependent ligand-binding affinities or methylation-dependent changes in complex size, may also constitute potential candidates for explaining the apparent changes in the dose-response behavior for different methylation levels. However, these alternative models would require more (and fine-tuned) parameters. And as the above equations demonstrate, a simple two-state model with constant binding affinities is sufficient to explain the phenomenon very well. Simple models are generally preferred over more complicated models, unless there is data contradicting the simple model.

How does the MWC model describe signaling by multiple receptor types in a complex? In this case, the energy difference is again given by the sum of the energy differences from all the constituent receptors. For two types of receptors, such as for $N_a$ Tar and $N_s$ Tsr receptors per complex with the total number of receptors given by $N = N_a + N_s$, the total energy difference is given by

$$\Delta F = N\Delta E + N_a k_B T \ln \left( \frac{1 + c/K_a^{\text{off}}}{1 + c/K_a^{\text{on}}} \right) +$$
$$N_s k_B T \ln \left( \frac{1 + c/K_s^{\text{off}}}{1 + c/K_s^{\text{on}}} \right). \tag{5.27}$$

Hence, even if only Tar receptors bind ligand (assuming the ligand-dissociation constants for the Tsr receptor are very large for a Tar-specific ligand), the energy difference of the whole complex is changed due to the first term in Eq. (5.27). Specifically, all receptors—those which bind and also those which do not bind—become equally methylated.

To demonstrate the effects of receptor cooperativity in Regime II for multiple receptor types, Fig. 5.6 shows dose-response curves for complexes of 1 and $N$ homogeneous receptors (Tar receptor), as well as for 2 and $2N$ mixed receptors of two different types (equal amounts of Tar and Tsr receptors). Clearly, the more receptors per complex the steeper the curve for highly methylated receptors. Also evident is the plateau development for mixed complexes. The reasons for this is that at low attractant, only Tar receptors bind and the Tsr receptors prevent the whole complex from turning off. Only when the attractant concentration is high enough, also the Tsr receptors start binding ligand and turn the whole complex off. This biphasic response explains the data in Fig. 2.7B.

The model for mixed-receptor complexes also provides a natural way for integrating different signals in presence of multiple types of chemical molecules. In the model, different receptor types form complexes with average receptor fractions corresponding to the abundances of the different receptor types in the cell. Each receptor type can bind ligand molecules with their characteristic binding affinities, and hence contributes to the energy difference of the complex. As illustrated in Fig. 5.7, signals are integrated so that there is only one complex activity, which can be read out by the receptor associated kinase CheA.

The design of funneling multiple stimuli into a single activity for a complex leads to a number of interesting consequences. The largest

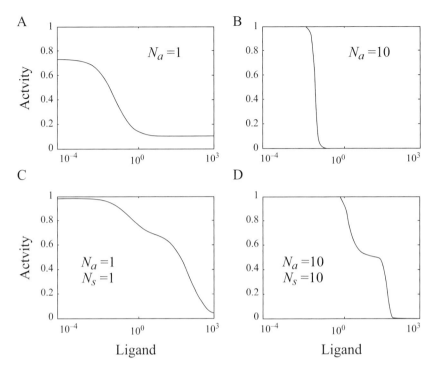

**Fig. 5.6** Dose-response curves in Regime II illustrating the effects of receptor complex size and hence cooperativity. (A) Single receptor and (B) homogeneous receptor complex (Tar). (C) Dimer and (D) complex of two mixed receptor types (Tar and Tsr). Based on Ref. (6).

determining factor of the energy difference is either given by the largest ligand-binding fraction of receptors or by the fraction of receptors which binds ligand most strongly. Due to this effect, favorable and unfavorable signals can be averaged out. A repellent can be compensated by an attractant, rendering its dangerous effect undetected. Here it clearly becomes important that cells have additional sensing mechanism and cross-talk with other signaling pathways. Especially, the regulation of individual receptor-type expression levels is an important mechanism for the cell to further control its chemotactic preferences.

## 5.5  Alternative Ising model for receptor cluster

In theory, one can imagine alternative receptor cluster models and ask how they compare with the data. Most important is the so-called Ising lattice, originally used to describe magnetism (Fig. 5.8). In this model, atoms are described by their magnetic spins, little vectors similar to compass needles describing the intrinsic magnetic properties of the atoms. These spins can align with the externally applied magnetic field. If one is interested in the magnetic property of the material in some spatial direction, e.g. the $z$-axis of a coordinate system, spins can either point up or down. Spins between neighboring atoms can also couple leading to

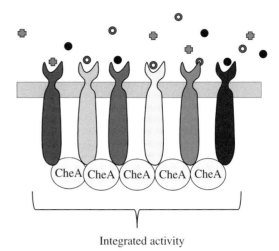

**Fig. 5.7** The energy contributions from different receptors in a complex are integrated into a single energy difference and hence a single activity of the whole complex.

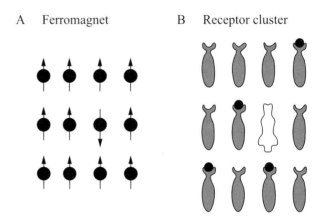

**Fig. 5.8** Lattices of ferromagnetic spins (A) and coupled receptors (B). Based on idea in Ref. (3).

collective, long-range behavior. Specifically in a ferromagnetic material, spins between neighboring atoms prefer to align parallel (Fig. 5.8A).

To illustrate the ferromagnetic Ising lattice, we write down the total energy of the lattice, also called the Hamiltonian, using the notation that $\sigma_i = +1$ and $\sigma_i = -1$ represent spin up and spin down, respectively. As a result, the Hamiltonian is given by

$$\Delta F = \frac{1}{2} \sum_i \Delta f \sigma_i - J \sum_{\langle i,j \rangle} \sigma_i \sigma_j \qquad (5.28)$$

with the sum over all lattice sites, $f = -2\mu B$ the magnetic-field splitting, and $J > 0$ the ferromagnetic coupling. All energies are in units of $k_B T$. Brackets $\langle ... \rangle$ indicate that double sum is restricted over neighboring pairs of sites on the lattice.

For an infinitely long 1D Ising chain, the magnetization can be calculated analytically and is given by

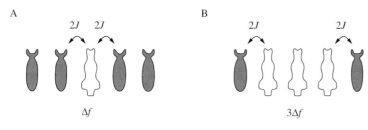

Single-receptor excitation                    Triple-receptor excitation

Fig. **5.9** Two representative excitations of the receptor activity in a 1D lattice with energetic costs as indicated. The single receptor excitation (A) is exponentially more favorable in energy compared to a triple receptor excitation (B).

$$M = \langle \sigma \rangle = \frac{1}{2}\left(1 - \frac{\sinh(\Delta f/2)}{\sqrt{e^{-4J} + \sinh^2(\Delta f/2)}}\right) \tag{5.29}$$

with $\sinh(x) = (e^x - e^{-x})/2$ the hyperbolic sine.

A similar Hamiltonian can be constructed for two-state receptors on a lattice, where each receptor is characterized by energy difference $\Delta f$ (Eq. (5.16)) and neighboring receptors are "ferromagnetically" coupled to favor the same activity state. Spin up and down correspond to active and inactive receptors, respectively (see Fig. 2.8B). The average receptor activity is calculated as $A = (M+1)/2$ with $M$ given by Eq. (5.29) in 1D. A delicate point is the so-called phase transition in the 2D square lattice, where each receptor has four equivalent neighboring receptors. For larger and larger coupling constant $J$, one reaches a critical coupling constant, $J_c = \ln(\sqrt{2} + 1)/2 \approx 0.44$, above which receptors form spontaneously long-range correlations of the activity even if there is no energy difference ($\Delta f = 0$). This results in hysteresis, the phenomenon that the activity is history dependent. As this has not been observed experimentally, coupling constant would need to be chosen relatively small ($J < J_c$).

While the cooperative response of methylated receptors in Regime II is produced nicely by the lattice model, it cannot properly describe the high sensitivity of demethylated receptors in Regime I. The reason is that in the lattice model the small residual activity in this regime stems from individual receptors, not collective receptor entities, as it is energetically more favorable to turn on individual receptors than groups of several receptors. This is demonstrated in Fig. 5.9 for the 1D lattice, showing the costs of a single and a triple receptor excitation in the activity. The penalty in breaking the ferromagnetic arrangement is the same for both types of excitations, but the cost in the receptor energy difference scales linearly with the number of flipped, active receptors. As the energy difference enters the activity in form of exponential Boltzmann factors, large receptor excitations are extremely unlikely to occur. Hence, the high sensitivity through receptor clustering is missing in the receptor lattice model.

However, as many physical interactions are local, the Ising model provides valuable insight into receptor-receptor correlations and hence response times.[2] To see this, we write a kinetic version of the Ising lattice using the master equation

[2]Based on Ref. (9)

$$\frac{d}{dt}p(\sigma_1,\ldots,\sigma_N;t) = -\left(\sum_i k_i(\sigma_i)\right)p(\sigma_1,\ldots,\sigma_N;t)$$
$$+ \sum_i k_i(-\sigma_i)p(\sigma_1,\ldots,-\sigma_i,\ldots,\sigma_N;t), \quad (5.30)$$

which describes the dynamics of the joint probability of receptor occupation $p(\sigma_1,\ldots,\sigma_N;t)$, assuming ligand binding and unbinding is much faster than receptor switching between *on* and *off* states. To describe the receptor switching rates we use the so-called *Glauber dynamics*

$$k_i(\sigma_i) = \frac{1}{2}\alpha\left(1 - \frac{1}{2}\gamma\sigma_i(\sigma_{i-1}+\sigma_{i+1})\right)(1-\beta\sigma_i) \quad (5.31)$$

with $\alpha$ the intrinsic switching rate of a single receptor, $\gamma = \tanh(2J)$, and $\beta = -\tanh(\Delta f/2)$. Eq. (5.31) describes a rate proportional to the intrinsic rate, reduced if neighboring receptors are in the same state or increased if in the opposite state, and modulated by the receptor energy, which acts as an external magnetic field on a spin. With these ingredients, the correlation time among receptors of a 1D Ising chain with periodic boundary conditions (a "ring")[3] can be calculated. The correlation time is given by

[3]Periodic boundary conditions mean $\sigma_{N+1} = \sigma_1$.

$$\tau_c = \frac{1}{\alpha[1-\tanh(2J)]} \to \frac{1}{\alpha}e^{2J} \quad \text{for } J \gg 1, \quad (5.32)$$

which increases exponentially with $J$ due to the phenomenon of *critical slowing down*. Eq. (5.32) is valid for Ising chains shorter than the correlation length of receptors. Using scaling arguments, these results can be extended to higher dimensions, including the 2D lattice. This shows that cooperativity implies constraints on dynamics, or that amplification comes at the expensive of slow response times. However, it is possible that biology found ways to implement some degree of non-local receptor–receptor interactions to ease this restriction.

# Chapter summary

- Cooperativity, implemented by allostery or other mechanisms, is widespread in biological sensing and regulatory systems. Examples include hemoglobin and chemo-receptor complexes, both of which can be described by a MWC-type model.
- In the MWC model for receptor complexes, all receptors are either *on* (active) or *off* (inactive) together. The activity of a complex of $N_a$ Tar and $N_s$ Tsr receptors (and hence a total receptor number $N = N_a + N_s$) is the Boltzmann probability for the complex to be *on*, given by

$$p_{\text{on}} = \frac{1}{1 + e^{\Delta F/k_B T}} \qquad (5.33)$$

$$\Delta F(m, c) = F_{\text{on}} - F_{\text{off}} \qquad (5.34)$$

$$= N\Delta E + N_a k_B T \ln\left(\frac{1 + c/K_a^{\text{off}}}{1 + c/K_a^{\text{on}}}\right)$$

$$+ N_s k_B T \ln\left(\frac{1 + c/K_s^{\text{off}}}{1 + c/K_s^{\text{on}}}\right). \qquad (5.35)$$

with $K_a^{\text{on}}$ ($K_s^{\text{on}}$) and $K_a^{\text{off}}$ ($K_s^{\text{off}}$) indicating the ligand dissociation constants of the Tar (Tsr) receptor in the *on* and *off* states, respectively. Hence, this energy difference of the complex is the sum of all individual receptor energy differences, depending on both ligand concentration $c$ and average receptor methylation level $m$.

- The model predicts two different regimes for signaling. Demethylated receptors are in Regime I, characterized by their exponentially small activity and heightened sensitivity. Methylated receptors are in Regime II, characterized by their high, near maximal activity and strong cooperativity.

- An alternative model is the Ising lattice, originally used to describe magnetism. Although conceptually appealing, especially with respect to describing realistic response times, this model has difficulties to describe signaling in Regime I.

# Further reading

(1) 'On the nature of allosteric transitions: a plausible model', by Jacques Monod, Jeffries Wyman, and Jean-Pierre Changeux, *J. Mol. Biol.* 12: 88–118 (1965).

(2) 'Temporal comparisons in bacterial chemotaxis', by Jeffrey E. Segall, Steven M. Block, Howard C. Berg, *Proc. Natl. Acad. Sci. USA* 83: 8987–8991 (1986).

(3) 'Receptor clustering as a mechanism to control sensitivity', by Dennis Bray, Matthew D. Levin, Carl J. Morton-Firth, *Nature* 393: 85–8 (1998).

(4) 'Heightened sensitivity of a lattice of membrane receptors', by Tom A.J. Duke and Dennis Bray, *Proc. Natl. Acad. Sci. USA* 96: 10104-10108 (1999).

(5) 'Quantitative modeling of sensitivity in bacterial chemotaxis: the role of coupling among different chemoreceptor species', by Bernardo A. Mello and Yuhai Tu, *Proc. Natl. Acad. Sci. USA* 100: 8223–8228 (2003).

(6) 'Chemosensing in Escherichia coli: two regimes of two-state receptors', by Juan E. Keymer, Robert G. Endres, Monica Skoge, Yigal Meir, and Ned S, Wingreen, Proc. *Natl. Acad. Sci. USA* 103: 1786–1791 (2006).

(7) *Scaling and Renormalization in Statistical Physics*, by John Cardy (Cambridge lecture notes in physics) (Cambridge University Press, 1996).

(8) 'Receptor–receptor coupling in bacterial chemotaxis: evidence for strongly coupled clusters', by Monica Skoge, Robert E. Endres, and Ned S. Wingreen, *Biophys. J.* 90: 4317–4326 (2006).

(9) 'Dynamics of cooperativity in chemical sensing among cell-surface receptors', by Monica Skoge, Yigal Meir, and Ned S. Wingreen, *Phys. Rev. Lett.* 107: 178101 (2011).

# 6

# Robust precise adaptation

For chemotaxis to work over a wide range of chemical concentrations, the cell requires that its receptors adapt to the current chemical concentration in the environment. If receptors do not adapt, their response would quickly saturate to zero or maximal activity when the cell swims in a chemical concentration gradient. This would make the cell "blind" to any further changes in the chemical concentration. Through adaptation the receptor activity stays near the most sensitive point at half-maximal activity, ready to respond. As a consequence, adaptation makes responses to persistent stimulation (such as a constant change in ligand concentration) only a transient phenomenon.

In Chapter 2 we described how adaptation is biochemically implemented by receptor methylation through CheR, mainly at four glutamate residues, this compensates for increases (decreases) in attractant (repellent) concentration and increases the activity back to the prestimulus value. In contrast, receptor demethylation by phosphorylated CheB (CheB-P) decreases the activity when cells encounter a decrease (increase) in attractant (repellent) concentration. Remarkably, as indicated in Figs. 2.5 and 2.6, adaptation to persistent stimulation is very precise. The adapted activity from FRET is largely independent of the ligand concentration. Furthermore, the precision of adaptation was found to be robust in experiments, as tempering with the expression level of chemotaxis proteins only affects the level of the adapted activity or the time it takes to adapt, not the precision.

As the process of adding or removing methyl groups is rather slow (compared to excitatory signaling), the methylation level is a kind of memory of previous ligand concentrations for the cell. However, as it is not read out directly by any known process, adaptation can also be regarded as the process of erasing information on current external conditions, encoded by the excitation. If adaptation is fast, cells stop responding and erase information very quickly, while if adaptation is slow, cells remember current conditions for a long time. Note that slow processes are generally suitable mechanisms for implementing memory in a cell. Using this definition, the amount of protein in a cell can be regarded as memory too.

## 6.1 Energy-landscape picture of adaptation

In Chapter 5 we introduced the two-state model for receptor complexes. The activity only depends on the energy difference $\Delta F$ between the *on* and *off* state of the complex. We learned that this energy difference in return depends implicitly on the ligand concentration $c$ and receptor-methylation level $m$. As summarized in Fig. 6.1, attractant turns receptors off by increasing $\Delta F$. In response, methylation lowers the on state and turns receptors back on by decreasing $\Delta F$. The only problem with this mechanism is to explain how the enzymes "know" when to stop methylating or demethylating. One way to achieve this is that the rates of methylation and demethylation are fine tuned to exactly compensate the effects of ligand concentration. But this would not work in a variety of different ligand concentration as the rates would need to change again.

## 6.2 Dynamics of adaptation

Adaptation can also be formulated in the language of dynamics. We could be tempted to write down a simple mass-action equation, which would describe the rate of change of the average receptor methylation level in a complex, e.g.

$$\frac{dm}{dt} = k_R \cdot R \cdot c \cdot (m_{\max} - m) \; - \; k_B \cdot B \cdot A(c, m) \cdot m. \qquad (6.1)$$

In this equation, the rate of methylation (the first term on the right-hand side) is proportional to the CheR concentration $R$, external ligand concentration $c$, and available methylation sites $m_{\max} - m$. The rate of demethylation (second term on right-hand side) is proportional to the

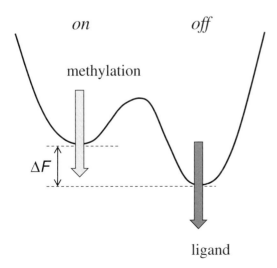

**Fig. 6.1** Energy landscape of the two-state model for a receptor complex. The energy difference $\Delta F$ determines the probability to be active. Also shown are the effects of ligand (attractant) and methylation on *off* and *on* state energies.

active CheB concentration (here taken to be the CheB concentration $B$ times the activity $A$) and methylated sites $m$. At steady state, $dm/dt = 0$ and the resulting equation can formally be solved for methylation level $\bar{m}$. Subsequently, the adapted activity $\bar{A}$ can be calculated. However, there is no reason to believe that this adapted activity is the same as the activity before the stimulus was applied (unless by accident or fine tuning of rate constants), and hence this adaptation model does not lead to robust precise adaptation.

Instead of fine tuning, the system can be designed in a specific way to create the property of robust precise adaptation inherently. In fact, the two-state model for signaling lends itself to an elegant way for achieving this remarkable feature. If the dynamics of receptor methylation and demethylation depend only on the activity of the receptors (and not on the number of available sites for methylation and demethylation, or the external ligand concentration), adaptation will always make the activity go back to the same adapted activity. For instance, instead of Eq. (6.1) the dynamics may be given by

$$\frac{dm}{dt} = k_R \cdot R \cdot (1 - A) \; - \; k_B \cdot B \cdot A, \tag{6.2}$$

where only inactive receptors become methylated and active receptors become demethylated. Evidently, Eq. (6.2) depends only implicitly on the ligand concentration and methylation level via $A$. At steady state, the resulting equation $0 = k_R R(1 - \bar{A}) - k_B B \bar{A}$ can easily be solved for the adapted activity

$$\bar{A} = \frac{k_R R}{k_R R + k_B B}, \tag{6.3}$$

which only depends on intrinsic rate parameters and enzyme concentrations, and hence is independent of the external conditions of the cell.

In other words, the adapted activity is a stable fixed point of the system. This can be illustrated by plotting the rates of methylation and demethylation as a function of the activity in Fig. 6.2. The intersection is the fixed point at which both rates exactly balance one another, and there is no net change in the receptor methylation level; the receptors are adapted to activity $\bar{A}$. When there is a perturbation, e.g. addition

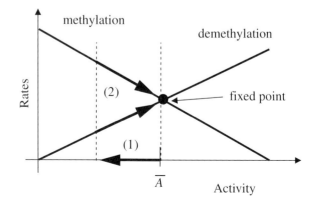

**Fig. 6.2** Flow diagram of adaptation with activity-dependent rates for methylation and demethylation. Also shown are the fixed point, corresponding to the adapted activity, and the dynamics (1) and (2) in response to an increase in attractant concentration. Based on Ref. (1).

of attractant, the activity drops shown by path (1). Subsequently, the rate of methylation exceeds the rate of demethylation, and naturally, the dynamical system flows back to the fixed point following path (2). In this case, the fixed point is said to be stable. (If it were unstable, the deviation of the activity from the adapted value would instead increase further and further.)

While Eq. (6.2) leads to precise adaptation for the cell, the level of the adapted activity may well be different from cell to cell as expression levels of receptors and enzymes (CheR and CheB) may vary. Similarly, the time to adapt and to return back to the adapted activity will vary from cell to cell. This can be easily seen from the relaxation rate $\lambda = k_R R + k_B B$, extracted from the linearized dynamics. Introducing small deviations from the adapted activity, activity error $\delta A = A - \bar{A}$, and small deviations from the original adapted methylation level $\delta m = m - \bar{m}$, the new dynamics become

$$\frac{d(\delta m)}{dt} = -\lambda \delta A. \tag{6.4}$$

Again the relaxation rate $\lambda$ depends only on cell-specific rate constants, not on cell-external parameters.

## 6.3   Chemotactic response function

The linearized model can be used to further characterize precise adaptation. In Chapter 3 we briefly introduced the linear response to a perturbation. Here, let us consider $R(t)$ as the response in bias in the fraction of time spend in the running mode (CCW motor rotation) to a change in ligand concentration $c(t)$ via

$$R(t) = \int_{-\infty}^{t} \chi(t - t')c(t')dt', \tag{6.5}$$

where $\chi(t - t')$ is the linear response function. Chemotactic bacteria constantly update their current probability of running or tumbling by differentially weighting the very recent measurement of ligand concentration against a measurement from the more distant past (a few seconds ago). This weighting is done by the chemotactic response function $\chi(t)$. In particular, by using a specific form of the response function, bacteria can efficiently bias their movement up chemical gradients of attractants.

The response function can be determined from applying a short ligand pulse to chemotactic bacteria and observing their swimming or motor behavior. Mathematically, applying sharp ligand pulse at time zero, $c(t) = \Delta c_t \delta(t)$ with $\Delta c_t$ a constant with dimension concentration times time, Eq. (6.5) becomes

$$R(t) = \Delta c_t \chi(t) \sim \chi(t). \tag{6.6}$$

Hence, the response function can directly be determined from the response itself.

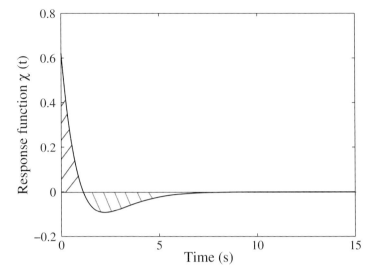

**Fig. 6.3** Chemotactic response function $R(t)$. The dark-shaded positive and the light-shaded negative lobes of the response have an approximately equal area, indicative of precise adaptation. The calculated curve agrees well with actual data from Ref. (5).

A theoretically derived response function is shown in Fig. 6.3 in good agreement with actual data. Visibly there is a positive lobe at early times (recent past) and a negative lobe at later times (more distant past), allowing the cell to determine changes in ligand concentration. If the integral in Eq. 6.5 is positive the run duration is extended, if it is negative the run duration is shortened.

In Chapter 9 we will derive this response function in more detail. However, irrespective of how to obtain the response function, here we want to note already that precise adaptation requires that the total area under the response function to be exactly zero, at least within this linear theory. To see this, assume that the response function was experimentally determined for the upstream receptor activity. Let us now integrate Eq. (6.4) for the deviation of activity from the adapted activity

$$\Delta m = \int \delta m(t)dt = -\lambda \int \delta A(t)dt, \qquad (6.7)$$

where $\delta A(t)$ equals the response function $\chi(t)$ for a ligand pulse. Since after the ligand pulse is over no net change in methylation level could have occurred, $\Delta m = 0$. (Although methylation started during the pulse, after the pulse these are removed again as the change in ligand concentration has been removed.) As a result, the integral of the response function is indeed zero. Experimentally, the area was found to be zero for some attractants such as aspartate, but not for other attractants such as serine.

## 6.4    Integral-feedback control

There is yet another way of looking at the problem of robust precise adaptation. The development of a control system that robustly tracks a specific steady-state output value is a problem often faced in engineering. A common solution is integral feedback control, in which the time integral of the error—the difference between the actual output and the desired steady-state output—is fed back into the system. This type of control ensures that the steady-state error approaches zero despite fluctuations in the input or in the system parameters. Man-made examples of integral feedback control include complex systems such as modern jet airplanes, refineries, power-grid stations, computer servers, as well as simple examples such as the thermostat.

One way to implement a thermostat is to check the current room or oven temperature and, if too low, to add some heat $\Delta Q \sim \Delta T$, where $\Delta T$ is the current error in temperature. As there will be unavoidable fluctuations of the temperature, some of these additions of heat will be unnecessary or even unwanted. As a result, the temperature starts to fluctuate even more. If instead we add heat proportional to the integrated error, such as

$$\Delta Q \sim \int \Delta T(t) dt, \tag{6.8}$$

small fluctuations in the temperature are evened out and hence do not negatively affect the stability of the temperature.

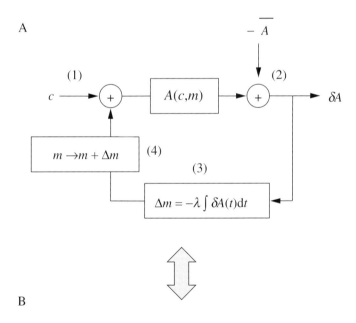

**Fig. 6.4** Control theory approach to adaptation. (A) Schematic of integral feedback control often used in engineering. Based on Ref. (6). (B) Equivalent representation using a system which combines a differentiator and a first-order low-pass filter in series. Based on Ref. (7).

The principle of integral-feedback control can also be found in our model for precise adaptation, Eq. (6.2). Integration of Eq. (6.4) for the error $\delta A(t)$ leads to Eq. (6.7), in which the number of added or removed methyl groups $\Delta m$ corresponds to the integrated error in the activity. Hence methylation does not correspond to the current error of the activity but the integrated error. This mechanism can be illustrated with a block-diagrammatic presentation from control theory. Fig. 6.4A illustrates the different steps in the algorithm, e.g. as if implemented on an analog computer: (1) input of ligand, (2) calculation of the error, (3) integration in time, and (4) change in number of methyl groups with corresponding change in activity.

Let us stay with the control engineering approach a little longer. It turns out that a system using integral feedback control is functionally equivalent to a system which uses a so-called *differentiator* and a *first-order low-pass filter* in series. While filters are introduced in more detail in Chapter 9, their concepts can be grasped by looking at Fig. 6.4B. A differentiator takes the derivative, here of the ligand concentration, allowing the cell to determine if the ligand concentration increases ($dc/dt > 0$) or decreases ($dc/dt < 0$). A first-order low-pass filter represents the time-dependent ligand concentration in terms of frequency $\omega$. As evident from the equation in the box, high frequencies with $\omega > K$ are filtered out.[1] In fact, the optimal cut-off frequency $K$ is the well-known Kalman filter from estimation theory. This filter is often applied in various engineering problems when one tries to remove high-frequency noise from the low-frequency signal.

[1]It is a first-order filter because the denominator is linear in $\omega$.

## 6.5    Assistance neighborhoods

The integral-feedback model of Eq. (6.2) based on the two-state receptor assumption solves the problem of robust precise adaptation elegantly. However, when adaptation of receptor complexes is simulated by stochastic approaches, such as the Gillespie algorithm (see Appendix A), adaptation of receptor complexes is far from precise. The reason for this is illustrated in Fig. 6.5A. All receptors of a complex have the same activity. However, the methylation level can vary from receptor to receptor. This is because the addition or removal of methyl groups is a random process where receptors are chosen randomly for methylation or demethylation. As a result, it can happen that a CheR enzyme tries to methylate a fully methylated receptor, which does not work. Or it could happen that a CheB enzyme tries to demethylate a receptor which is already fully demethylated. This also cannot happen. Consequently, the rates of methylation and demethylation become dependent on the actual methylation level of the receptors, which leads to imprecise adaptation.

The problem of imprecise adaptation for receptor complexes can be remedied if an additional experimental observation is included in the stochastic simulation: enzymes actually do not freely diffuse, but tether

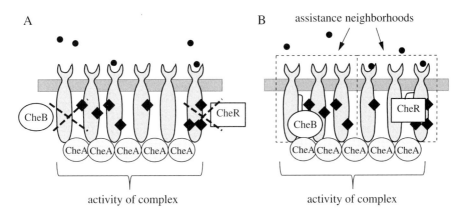

**Fig. 6.5** Adaptation of a receptor complex without (A) and with (B) "assistance neighborhoods" (dashed boxes). The former can lead to aborted methylation and demethylation attempts (crosses), which causes imprecise adaptation. Based on Ref. (8).

to the receptors (to the flexible C-termini of the receptor proteins) to act upon groups of about 6–8 receptors in their vicinity. These groups are called "assistance neighborhoods", and are illustrated in Fig. 6.5B. In effect, this boils down to each enzyme "seeing" a large set of potential modification sites to act upon for methylation or demethylation. Hence, the group of receptors is almost like a single large receptor with many modification sites, avoiding the problem of the "end points" of fully methylated or demethylated receptors where further modification can break down. As a result, the activity of the complex correlates much more strongly with the average receptor methylation level. Precise adaptation for cooperative receptors is robustly restored.

Based on these and other observations, we slightly modify our adaptation model, Eq. (6.2), and formally think of the dynamics of methylation in terms of the total methylation level $M$ of the whole receptor complex

$$\frac{dM}{dt} = N_R(R) \cdot g_R \cdot (1 - A) \ - \ N_B(B_p) \cdot g_B \cdot B_p^n \cdot A \qquad (6.9)$$

with $M = Nm$ and $N$ is the total number of receptors in the complex. The other changes in Eq. (6.9) are as follows: first, due to the assistance neighborhoods the rate of methylation depends on the number of tethered CheR enzymes $N_R$ (assumed proportional to the CheR concentration $R$ so that $g_R$ is the rate constant of methylation per tethered CheR enzyme). Similarly, the rate of demethylation depends on the number of tethered CheB-P enzymes, $N_B$ (assumed also proportional to the CheB-P concentration so that $g_B$ is the rate constant of demethylation per tethered CheB-P enzyme). Second, we generalize the expression for the rate of demethylation by including an additional dependence on $n$th power of the CheB-P concentration, $B_p^n$, thus allowing the tethered CheB-P enzymes to cooperate with additional CheB-P enzymes in their vicinity (if $n > 0$). Eq. (6.9), or simplified versions of it, will be used for the dynamics of adaptation in the remainder of the book.

# Chapter summary

- *E. coli* cells adapt precisely to persistent stimulation, so that after a transient response the signaling activity and probability of tumbling always return to their basal levels.

- Based on the two-state property, Barkai and Leibler originally postulated a model for precise adaptation. In this model, the rates of methylation by CheR and the rate of demethylation by CheB only depend on the receptor activity, not the external ligand concentration or available receptor modification sites. This provides negative feedback, in addition to the negative feedback through CheB phosphorylation.

- In order for this model to work precisely for cooperative receptor complexes, the observation of assistance neighborhoods needs to be incorporated. This means that the adaptation enzymes tether to the flexible C-termini of the receptors to act on groups of receptors. This makes sure that the rates of modification do not depend on the number of available modification sites.

# Further reading

(1) 'Robustness in simple biochemical networks', by Naama Barkai and Stanislav Leibler, Nature 387: 913–17 (1997).

(2) 'Robustness in bacterial chemotaxis', by Uri Alon, Michael G. Surette, Naama Barkai, Stanislav Leibler, *Nature* 397: 168–71 (1999).

(3) 'Design principles of a bacterial signalling network', by Markus Kollmann, Linda Lvdok, Kilian Bartholomé, Jens Timmer, Victor Sourjik, Nature 438: 504–7 (2005).

(4) 'The bacterial chemotactic response reflects a compromise between transient and steady-state behavior', by Damon A Clark and Lars C Grant, *Proc. Natl. Acad. Sci. USA* 102: 9150–5 (2005).

(5) 'Bacterial strategies for chemotaxis response', by Antonio Celani and Massimo Vergassola, *Proc. Natl. Acad. Sci.* 107: 1391–1396 (2010).

(6) 'Robust perfect adaptation in bacterial chemotaxis through integral feedback control', by Tau-Mu Yi, Yun Huang, Melvin I. Simon, John Doyle, *Proc. Natl. Acad. Sci. USA* 97: 4649–53 (2000).

(7) 'Optimal noise filtering in the chemotactic response of Escherichia coli', by Burton W. Andrews, Tau-Mu Yi, Pablo A. Iglesias, *PLoS Comp. Biol.* 2: e154 (2006).

(8) 'Precise adaptation in bacterial chemotaxis through assistance neighborhoods', by Robert G. Endres and Ned S. Wingreen, *Proc. Natl. Acad. Sci. USA* 103: 13040–4 (2006).

(9) 'Chemotactic response and adaptation dynamics in Escherichia coli', by Diana Clausznitzer, Olga Oleksiuk, Linda Lvdok, Victor Sourjik, Robert G. Endres, *PLoS Comput. Biol.* 6: e1000784 (2010).

# Polar receptor localization and clustering

Receptors of bacterial chemotaxis predominantly cluster at the cell poles, effectively turning the cell ends into a sensitive "nose" for chemical detection. Interestingly, this clustering occurs at multiple length scales. At the smallest scale, receptors of the same type, e.g. the Tar receptor, form dimers.[1] Subsequently, dimers assemble into stable trimers of dimers of mixed dimer type (see Fig. 7.1). At a large scale, trimers form large polar clusters of approximately 0.1-$\mu$m diameter, and composed of thousands of receptors.

In previous chapters we discussed the functional role of receptor clustering in terms of signal amplification and precise adaptation. However, how do receptors localize at the cell poles in the first place, and what factors determine the cluster size? In this chapter we discuss the physical principles of polar receptor clustering based on strongly coupled trimers of dimers. Cluster size and localization will be determined by a minimum of the cluster-membrane energy.

## 7.1 Trimer of dimers

Polar localization appears to be an intrinsic property of chemoreceptors (and hence may rely largely on physical principles). Experimental observations show that localization hardly depends on the presence or absence of the receptor-bound kinase CheA and adapter protein CheW, and is largely unaffected by removal of the periplasmic ligand-binding domain of the receptors. Polar localization is likely a passive (little regulated) process, since newly synthesized receptors, initially inserted at random positions in the membrane, diffuse and ultimately become trapped at the cell poles. Importantly, imaging experiments indicate that polar receptor localization may depend on membrane curvature. In particular, the experimentally determined (partial) trimer structure implies a high intrinsic curvature, which may lead to timers preferably localizing to curved membrane areas on the cell surface. Taken together, these observations suggest that the distinct trimer structure with its high intrinsic curvature holds the key for unraveling polar receptor localization.

To understand how clustering works, we first need to know about the membrane properties—in particular, of the inner membrane in *E. coli*. The membrane is made of a lipid bi-layer with the polar head groups at

[1] These homo-dimers are what we have been calling "receptors" so far since they can bind maximally one ligand molecule.

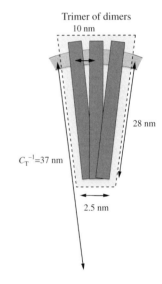

**Fig. 7.1** Geometry of trimer of dimers, embedded in the inner membrane of *E. coli*. The intrinsic curvature of a trimer is denoted by $C_T$. Redrawn from Ref. (1).

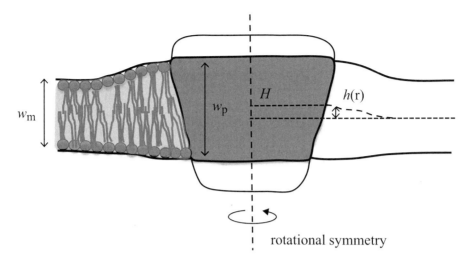

**Fig. 7.2** Membrane-embedded solid protein deforms the surrounding liquid membrane. The membrane with equilibrium width $w_m$ needs to deform to match the width $w_p$ of the hydrophobic region of the protein. The shape of the membrane relative to the equilibrium shape is given by $h(r)$, where $r$ is the radial distance from the center of the protein. Furthermore, $H$ is the height of the deformation at the interface of cluster and membrane. Note that rotational symmetry around the center of the protein is assumed to aid the analysis. Based on Ref. (3).

the outside to face the polar water molecules. The non-polar, hydrophobic lipid tails are on the inside. Receptors, like other membrane-embedded proteins, have a hydrophobic region, which anchors the protein in the membrane (Fig. 7.2). If the width of the hydrophobic region $w_P$ is larger than the membrane thickness $w_m$, the membrane will expand to match the protein region. If the width of the hydrophobic region is smaller than the membrane thickness, the width of the lipid bi-layer will shrink to match the protein. Hence, the protein, simplified as a solid object here, deforms the membrane around it. This costs some bending and stretching energy, but is offset by the hydrophobic attraction between the hydrophobic lipid tails and hydrophobic protein region.

Since a trimer may induce high membrane curvature, as illustrated in Fig. 7.1, it is more energetically favorable for the trimer to be at the poles, which naturally have a high curvature, than to be at the lateral cell area with a low curvature. However, the energy gain for the localization of a single trimer is rather small, about a fraction of the thermal energy $k_B T$, and hence may not be sufficient for extensive clustering at the poles. As a result, tight clustering requires additional energetic coupling between the trimers (trimer–trimer attraction). While there is evidence for this coupling from the receptor cooperativity (Chapters 2 and 5), the actual mechanism of coupling is not understood. In fact, there are several potential candidates for such a mechanism (see Fig. 7.3). The conventional view is that CheA and CheW form a kind of molecular linker, which mediates the interactions between receptor trimers (panel A). Alternative models include swapping of the cytoplasmic receptor domains (panel B) and membrane-mediated coupling (panel C). The latter is a result of energy-costing membrane deformations around one trimer of dimer, which are reduced if trimers cluster. (This is because the deformation does not need to decay to the equilibrium membrane shape away from the trimer, but just needs to span to the next deformation from a nearby trimer.) Such membrane deformations can arise if the hydrophobic region of a trimer is larger or smaller than the width of the membrane.

## 7.2 Elastic cluster-membrane model

In the following we describe a simple model for polar receptor localization and clustering due to the high intrinsic curvature of trimers and an attractive trimer–trimer coupling. Specifically, we consider a membrane-embedded cluster composed of trimers as its basic building blocks. For a sphero-cylindrical cell with radius $R_{\text{cell}}$ (see Fig. 7.4), we assume that the average membrane curvature at the poles, $1/R_{\text{cell}}$, is twice as large as the average curvature at the lateral cell area, $1/(2R_{\text{cell}})$, and that trimers have an even higher intrinsic curvature. The intrinsic curvature of the trimers (and hence of the cluster) tends to deform the membrane, penalizing large clusters of trimers. However, attractive coupling between trimers drives cluster formation, leading to a competition between these two opposing energy contributions. Using continuum elastic theory, we can derive an analytical expression for the total cluster-membrane energy. We will demonstrate that, due to the reduced cluster-membrane curvature mismatch at the poles, trimers favorably cluster at the poles and not at the lateral cell area.

The total energy of the cluster-membrane system generally has three contributions. There are the elastic energies $E_{\text{clu}}$ and $E_{mem}$ of the cluster and surrounding membrane, respectively. The cluster and membrane have to match each other at the interface and may be deformed from their preferred shape. This is because the curvature of the membrane-embedded cluster may not exactly match the underlying curvature of the constituent trimers. Similarly, the curvature of the membrane, due to the link with the cluster, may not match the equilibrium-membrane shape. The equilibrium-membrane shape is determined by the turgor pressure, which pushes the membrane (and cluster) against the cell wall (Fig. 7.5). This pushing effectively leads to a pinning-energy contribution to the cluster and membrane energies, penalizing deviations from the equilibrium-cell shape. Furthermore, there is the attractive trimer-trimer

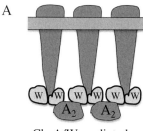

A

CheA/W mediated

B

Domain swapping

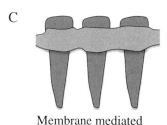

C

Membrane mediated

**Fig. 7.3** Potential trimer–trimer coupling mechanisms, mediated by CheA and CheW proteins (A), by domain swapping of cytoplasmic receptor domains (B), or by membrane deformations (C).

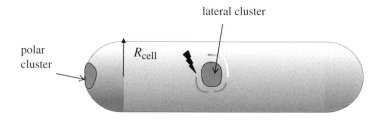

**Fig. 7.4** Sphero-cylindrical cell with a polar and a lateral receptor cluster. The radius of the cell cap and cylinder is given by $R_{\text{cell}}$. The flash symbol indicates energetically unfavorable membrane deformations.

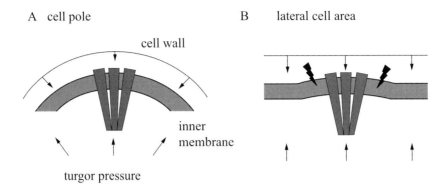

**Fig. 7.5** Cluster of three trimers at the cell pole (A) and at the lateral cell area (B). Effects of membrane curvature, turgor pressure, and pinning force from the constraints of the cell wall are shown. The flash symbols in (B) indicate energetically unfavorable membrane deformations. Redrawn from Ref. (1).

energy $E_a$, which drives cluster formation. Taken together, the total energy is given by

$$E = \underbrace{E_{\text{clu}} + E_{\text{mem}}}_{E_{\text{el}}} + E_{\text{att}}. \tag{7.1}$$

As illustrated in Fig. 7.5, the elastic energy $E_{\text{el}}$ is expected to be different if the cluster is at the pole or at the lateral cell area since an intrinsically curved cluster will deform the membrane more at the lateral cell area.

Before describing a fully elastic cluster, we initially calculate the deformation of the membrane around a solid cluster. To ease the calculation, we first approximate the curved membrane by a locally flat membrane, with any deviations in curvature representing deviations from the underlying curved membrane. In this approximation the cluster is circular and the surrounding membrane is a 2D, elastic sheet with rotational symmetry around the cluster center. Second, we make the simplifying assumption that the cluster radius is rather large (but still much smaller than the cell dimensions). That way we can neglect the 2D character of the problem and instead consider the approximate 1D problem. Once this is solved, we will approximate the 2D solution by wrapping the membrane around the cluster as indicated in Fig. 7.6.

In the 1D case, we use coordinate $x$ instead of radius $r$, and introduce the function variable $h(x)$ to measure the lipid-membrane deformation away from its (curved) equilibrium value (starting at value $H$ at $x = R$, see Fig. 7.2). With these preparations complete, we can write the energy of the membrane (actually energy per length) as

$$E_{\text{mem}}[h(x)] = \underbrace{\frac{\kappa}{2} \int_R^\infty \left(\frac{d^2h}{dx^2}\right)^2 dx}_{\text{bending energy}} + \underbrace{\frac{\lambda}{2} \int_R^\infty h(x)^2 dx}_{\text{pinning energy}}, \tag{7.2}$$

where notation $E[h(x)]$ indicates that the energy is a functional of $h(x)$. The first term on the right-hand side of Eq. (7.2) is the bending energy,

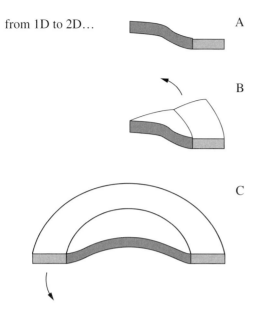

from 1D to 2D...

A

B

C

**Fig. 7.6** The 1D solution is used to approximate the full 2D solution for large clusters by wrapping the membrane around in a circle (A–C). Shown are a cluster (dark gray) with a thin surrounding membrane shell (light gray). Based on Ref. (4).

which depends on the membrane curvature (second derivative). The second term is the pinning energy of the membrane, which depends on the membrane height. (Taking their squared values allows these terms to conveniently become independent of their sign.)

In order to compute the energy associated with the membrane deformation, we must find the profile $h(x)$ that minimizes this energy subject to the boundary conditions.[2] To do this we will have to specify the value of $h(x)$ at the boundaries $x = R$ and for $x \to \infty$. At $r = R$ we assume strong hydrophobic "gluing"

$$h(R) = H, \tag{7.3}$$

and that the derivative $h' = \partial h / \partial x$ is (for simplicity) zero, i.e.

$$h'(R) = 0. \tag{7.4}$$

Furthermore, we make the realistic assumption that for $x \to \infty$ the membrane becomes the unperturbed membrane

$$h(\infty) = 0 \tag{7.5}$$

$$h'(\infty) = 0. \tag{7.6}$$

Minimizing the energy functional is analogous to minimizing an ordinary function by setting the derivative equal to zero. Here we set the functional derivative to zero and obtain a partial differential equation. To compute the function derivative, we let the function $h(x)$ undergo a small excursion $\eta(x)$ by substituting $h(x) \to h(x) + \epsilon \eta(x)$ with the boundary conditions on $h$ such that $\eta$ vanishes at the boundaries (see Appendix A). The functional derivative is defined as

[2]See Ref. (2) for a similar treatment of a channel protein.

$$\frac{\partial E}{\partial h(x)} = \lim_{\epsilon \to 0} \frac{E[h(x) + \epsilon\eta(x), h''(x) + \epsilon\eta''(x)] - E[h(x), h''(x)]}{\epsilon}. \quad (7.7)$$

Abbreviating the integrand in Eq. (7.2) by $f$, after Taylor expansion up to second order

$$f[h + \epsilon\eta, h'' + \epsilon\eta''] \approx f[h, h''] + \epsilon\frac{\partial f}{\partial h}\eta + \epsilon\frac{\partial f}{\partial h''}\eta'', \quad (7.8)$$

we obtain for the functional derivative (with the first term of Eq. (7.8) being canceled in Eq. (7.7))

$$\frac{\partial E}{\partial h(x)} = \epsilon\int_R^\infty \left(\frac{\partial f}{\partial h''}\eta'' + \frac{\partial f}{\partial h}\eta\right) dx = 0 \quad (7.9)$$

$$\to \kappa\int_R^\infty h''\eta'' dx + \lambda\int_R^\infty h\eta\, dx = 0. \quad (7.10)$$

We proceed by integrating the first term "by parts", resulting in

$$\kappa h''\eta'|_R^\infty - \kappa\int_R^\infty h'''\eta' dx + \lambda\int_R^\infty h\eta\, dx = 0. \quad (7.11)$$

The first term is zero as $\eta'$ vanishes at the boundaries. We can integrate by parts once more and obtain

$$-\kappa h'''\eta|_R^\infty + \kappa\int_R^\infty h''''\eta dx + \lambda\int_R^\infty h\eta\, dx = 0, \quad (7.12)$$

where the first term vanishes again. Since this integral is zero for any $\eta$, we obtain the fourth-order differential equation

$$\kappa\frac{d^4h}{dx^4} + \lambda h = 0 \quad (7.13)$$

with its solution delivering the membrane profile which minimizes the energy of the membrane around the cluster. To solve Eq. (7.13), we make an *Ansatz* of the form $h = e^{\Lambda x}$, leading to

$$\Lambda^4 + \frac{\lambda}{\kappa} = 0 \quad (7.14)$$

and

$$\Lambda = \sqrt[4]{\frac{\lambda}{\kappa}} \times \sqrt[4]{-1}. \quad (7.15)$$

The fourth root of $-1$ can easily be calculated by using the Euler formula $e^{i\theta} = \cos\theta + i\sin\theta$, i.e. $e^{i\pi} = -1$ in our case, and realizing that there are an infinite number of solutions $e^{i(2n+1)\pi} = -1$ for $n = 0, 1, 2, 3, \ldots$. However, the first four distinct solutions lead to angles $\pi/4$, $3\pi/4$, $5\pi/4$, and $7\pi/4$, and

$$\Lambda_1 = \sqrt[4]{\frac{\lambda}{\kappa}} e^{i\pi/4} = \sqrt[4]{\frac{\lambda}{\kappa}} \left( \frac{\sqrt{2}}{2} + i\frac{\sqrt{2}}{2} \right) \tag{7.16}$$

$$\Lambda_2 = \sqrt[4]{\frac{\lambda}{\kappa}} e^{i3\pi/4} = \sqrt[4]{\frac{\lambda}{\kappa}} \left( -\frac{\sqrt{2}}{2} + i\frac{\sqrt{2}}{2} \right) \tag{7.17}$$

$$\Lambda_3 = \sqrt[4]{\frac{\lambda}{\kappa}} e^{i5\pi/4} = \sqrt[4]{\frac{\lambda}{\kappa}} \left( -\frac{\sqrt{2}}{2} - i\frac{\sqrt{2}}{2} \right) \tag{7.18}$$

$$\Lambda_4 = \sqrt[4]{\frac{\lambda}{\kappa}} e^{i7\pi/4} = \sqrt[4]{\frac{\lambda}{\kappa}} \left( \frac{\sqrt{2}}{2} - i\frac{\sqrt{2}}{2} \right). \tag{7.19}$$

From these we choose the two solutions with negative real part, since only they decay to zero far away from the membrane-embedded cluster. Thus we obtain

$$h(x) = C_2 e^{\Lambda_2 x} + C_3 e^{\Lambda_3 x} \tag{7.20}$$

with constants $C_2$ and $C_3$ determined from the boundary conditions at $x = R$. Applying the boundary conditions leads to two algebraic equations

$$C_2 e^{\Lambda_2 R} + C_3 e^{\Lambda_3 R} = H \tag{7.21}$$

$$\Lambda_2 C_2 e^{\Lambda_2 R} + \Lambda_3 C_3 e^{\Lambda_3 R} = 0, \tag{7.22}$$

which can be solved and, plugged into Eq. (7.20), produce the solution

$$h(x) = H e^{-(\sqrt{2}/2)K(x-R)} \left[ \cos \frac{\sqrt{2}}{2} K(x-R) + \sin \frac{\sqrt{2}}{2} K(x-R) \right] \tag{7.23}$$

with $K = \sqrt[4]{\frac{\lambda}{\kappa}}$ the inverse decay length of the membrane perturbation.

To calculate the energy Eq. (7.2), we also need the second derivative

$$h''(x) = H K^2 e^{-(\sqrt{2}/2)K(x-R)} \left[ \sin \frac{\sqrt{2}}{2} K(x-R) - \cos \frac{\sqrt{2}}{2} K(x-R) \right] \tag{7.24}$$

and to solve the resulting integral

$$E_{\mathrm{mem}} =$$

$$\frac{K}{2} \int_R^\infty dx\, U^2 K^4 e^{-\sqrt{2}K(x-R)} \left[ 1 - 2\sin\frac{\sqrt{2}}{2} K(x-R) \times \cos\frac{\sqrt{2}}{2} K(x-R) \right]$$

$$+ \frac{K}{2} \int_R^\infty dx\, U^2 e^{-\sqrt{2}K(x-R)} \left[ 1 + 2\sin\frac{\sqrt{2}}{2} K(x-R) \times \cos\frac{\sqrt{2}}{2} K(x-R) \right]$$

$$= K H^2 \int_R^\infty dx\, e^{-\sqrt{2}K(x-R)} = \frac{\lambda H^2}{\sqrt{2}K}. \tag{7.25}$$

In addition to the costs of the deformation of the membrane, an incentive for clustering is needed. We assume that neighboring trimers interact by an attractive coupling energy $J$. Since this is a local interaction, the total attractive energy contribution $E_{\mathrm{att}}$ will be proportional to the number of trimers in the cluster (at least for sufficiently large clusters). Hence, in terms of the cluster radius, this energy is proportional to the square of the radius

$$E_{\mathrm{att}} \approx -J \cdot R^2 \tag{7.26}$$

with $J > 0$. Taken together, the total energy of the membrane is given by

$$E = E_0 + \underbrace{\frac{\lambda H^2}{\sqrt{2K}}}_{\text{energy/length}} \underbrace{2\pi R}_{\text{circumference}} - \underbrace{JR^2}_{\text{area}-\text{tension}}$$

$$= E_0 + \pi \bar{K} H^2 R - JR^2 \tag{7.27}$$

with $\bar{K} = \sqrt{2} \sqrt[4]{\lambda^3 K}$ and an (unimportant) additive constant $E_0$. Notice that the second term is positive and proportional to cluster radius $R$, hence describing a line tension. In contrast, the third term is negative and proportional to $R^2$, similar to an energy contribution from a surface tension. Consequently, the total energy is parabolic with respect to $R$, as shown in Fig. 7.7. While the parabolic function by itself would indicate that large clusters always form and are energetically favorable, this does not need to be the case for a finite number of trimers on the cell surface. For a relatively small number of trimers, large clusters may not form or are unfavorable due to the large entropic cost for bringing all trimers together into a cluster.

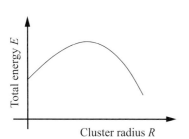

**Fig. 7.7** Parabolic cluster-membrane energy as a function of cluster radius. The attractive cluster energy is included, while the elastic energy of the cluster is not included yet.

## 7.3   Polar receptor clustering

We now return to the original question of why clusters from predominantly at the cell poles and not at the lateral areas of the cell. To get a grasp of this question we can write down an elastic energy for both the cluster and the membrane similar to Eq. (7.2). However, since we now have to start integrating at the origin, the 1D approximation will not work any more, so instead we integrate over the area using polar coordinates:

$$E_{el} = 2\pi \int_0^R \left[ \frac{\kappa_c}{2} (\nabla_r^2 h_c(r) - \Delta C)^2 + \frac{\lambda}{2} h_c^2(r) \right] r\, dr$$

$$+ 2\pi \int_R^\infty \left[ \frac{\kappa_m}{2} (\nabla_r^2 h_m(r))^2 + \frac{\lambda}{2} h_m^2(r) \right] r\, dr, \tag{7.28}$$

where the prefactor $2\pi$ comes from the integration of the rotation angle, the *Laplacian* is defined as $\nabla_r^2 = \partial^2/\partial r^2 + 1/r \, \partial/\partial r$, and "mismatch" $\Delta C$ is the preferred curvature of the cluster, $C_{\mathrm{TD}}$, minus the preferred curvature of the membrane (see below). We also introduced height functions $h_c(r)$ and $h_m(r)$ for the cluster and membrane deformations, and $\kappa_c$ and $\kappa_m$ for the specific bending stiffnesses of cluster and membrane, respectively.

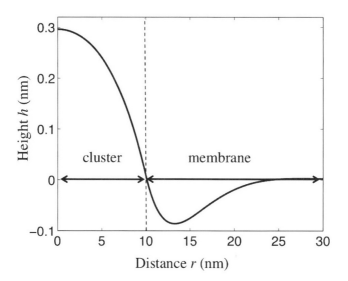

When the cluster and membrane deformations and finally total energy $E$ are calculated (similar to Eq. (7.27)), the linear term ($\sim R$) as well as the quadratic term ($\sim R^2$) obtain additional contributions due to the elasticity of the cluster. In simplified notation, we obtain

$$E = E_0 \; + \; cR \; - \; \underbrace{\left( J - \frac{1}{2}\pi\kappa_c\Delta C^2 \right)}_{\alpha} R^2 \qquad (7.29)$$

with coefficient $c$ not specified further here. The corresponding cluster-membrane deformation is shown in Fig. 7.8 for the full rotationally symmetric cluster.[3] One can clearly see that the profile fulfills the boundary conditions outlined above. Importantly, by inspecting Eq. (7.29) we can see that only for a sufficiently small mismatch between the preferred cluster and membrane curvatures ($\alpha > 0$), clustering is energetically beneficial. For large mismatch, the $R^2$ term becomes in fact positive ($\alpha < 0$) and no clusters form (Fig. 7.9A).

However, for a sphero-cylindrical cell, we have two different values for the membrane curvature - a high value at the poles and a low value at the lateral area. Hence, $\Delta C$ equals $C_{TD} - C_p$ for the pole and $C_{TD} - C_l$ for the lateral area. The larger the mismatch, the more the cluster (and with it the surrounding membrane) will deform from the cell's preferred equilibrium shape, and the larger the pinning-energy penalty. Since $C_{TD} > C_p > C_l$, clusters will preferably form at the cell poles over the lateral area.

To be more specific, the difference in total energy between a polar and a lateral cluster is

$$\Delta E = E_p - E_l = \frac{1}{2}\pi\kappa_c\left( C_{TD} - \frac{3C_p}{4} \right) C_p R^2, \qquad (7.30)$$

where we used that the curvature at the lateral cell area $C_l = C_p/2$ is half the curvature at the cell poles. This shows that there is a constraint

# Further reading

(1) 'Polar chemoreceptor clustering by coupled trimers of dimers', by Robert G. Endres, *Biophys. J.* 96: 453–63 (2009).

(2) *Physical Biology of the Cell*, by Rob Phillips, Jane Kondev, and Julie Theriot (Garland Science, 2004).

(3) 'Analytic models for mechanotransduction: gating a mechanosensitive channel', by Paul A. Wiggins and Rob Phillips, *Proc. Natl. Acad. Sci. USA* 101: 4071–6 (2004).

(4) *Role of Lipid Bilayer Mechanics in Mechanosensation*, by Tristan Ursell, Jane Kondev, Daniel Reeves, Paul A. Wiggins, and Rob Phillips, in *Mechanosensitive Ion Channels*, ed. Andre Kamkin and Irina Kiseleva (Springer, 2008).

(5) *Mechanics of the Cell*, by David Boal (Cambridge, 2003).

these estimates have an intrinsic uncertainty due to the randomness from molecule diffusion.

In the second part of this chapter we will extend the model to include the next step in actual sensing of ligand molecules by a cell, i.e. the effects of ligand-receptor binding on the accuracy of sensing. We will see that sensing by actual receptors cannot improve the accuracy of sensing. In fact, additional noise makes sensing less accurate.

## 8.1   Perfectly absorbing sphere

For the perfectly absorbing sphere shown in Fig. 8.3, the uncertainty in measuring chemical concentration is straightforward to derive. We start from the diffusion equation

$$\frac{\partial c}{\partial t} = D\nabla^2 c, \tag{8.1}$$

which governs the spatial and temporal variation of chemical concentration $c$. In Eq. (8.1), $D$ is the diffusion constant and $\nabla^2$ is the *Laplacian* (sum of second derivatives). At steady state, the left-hand side of Eq. (8.1) is zero, as the concentration does not change in time. As a consequence, the solution (in spherical symmetry) is $c(r) = \bar{c}(1 - a/r)$ with $c(r = a) = 0$ at $c(\infty) = \bar{c}$. The average molecule current, $Ddc/dr$, impinging on the sphere is $\bar{c}/r^2$, and integrated over the whole sphere, is given by $J = 4\pi Da\bar{c}$. Hence, the average number of molecules absorbed in time $\tau$ is $L_\tau = 4\pi Da\bar{c}\tau$. Since the molecules are independent, $L_\tau$ is Poisson distributed with the variance equal to the mean, i.e. $\langle(\delta L_\tau)^2\rangle = \langle L_\tau\rangle$, with brackets $\langle...\rangle$ indicating the average over an ensemble of many cells. Therefore, the perfectly absorbing sphere has a sensing uncertainty of

$$\frac{\langle(\delta c)^2\rangle_\tau}{\bar{c}^2} = \frac{\langle(\delta L_\tau)^2\rangle}{\langle L_\tau\rangle^2} = \frac{1}{4\pi Da\bar{c}\tau}. \tag{8.2}$$

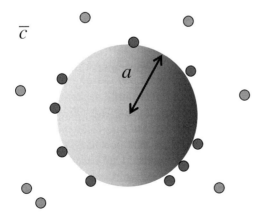

**Fig. 8.3** The perfectly absorbing sphere (model 1) estimates the external ligand concentration from the number of absorbed molecules on its cell surface. Redrawn from Ref. (4).

Hence, the measurement uncertainty depends only on cell-external parameters $D$ and $\bar{c}$, as well as "measurement-device" size $a$ and averaging time $\tau$. As expected, this uncertainty is the smaller the larger the measurement device (spatial averaging by $a$), the larger $D$, and the longer the measurement time (temporal averaging by $\tau$). While the absolute uncertainty $\langle(\delta c)^2\rangle$ increases with the average concentration $\bar{c}$, the relative uncertainty given by the ratio $\langle(\delta c)^2\rangle/\bar{c}^2$ decreases with increasing average concentration $\bar{c}$.

## 8.2   Perfectly monitoring sphere

[1]Based on *perfect instrument* in Ref. (2).

The perfectly monitoring sphere[1] is another parameter-free model for a cell that "perfectly" binds and releases all molecules that contact its surface (see Fig. 8.4). To quantify the time that a diffusing molecule spends in the cell's vicinity and is therefore capable of being measured, the cell is treated as a permeable sphere that infers the molecule concentration by counting the number of molecules $L$ inside its volume. The accuracy is further improved by averaging over several statistically independent measurements.

A simple estimate for the resulting uncertainty in concentration sensing is obtained as follows: the number $L$ is Poisson distributed, and the cell counts on average $\langle L\rangle = a^3 c$ molecules in its volume at any time. During a time $\tau$, the cell can make $N_m \approx \tau/(a^2/D)$ statistically independent measurements (see Fig. 8.5), with $a^2/D$ the typical turnover

[2]The turnover time is the time it takes to replace the molecules inside the cell by new molecules from outside the cell by diffusion.

time for molecules inside the sphere.[2] Taken together, this leads to

$$\frac{\langle(\delta c)^2\rangle_\tau}{\bar{c}^2} = \frac{\langle(\delta L)^2\rangle}{\langle L\rangle^2} = \frac{1}{N_m\langle L\rangle} \approx \frac{1}{Da\bar{c}\tau}. \tag{8.3}$$

This result depends on the same physical parameters as the result of the perfectly absorbing sphere, Eq. (8.2), although the prefactor is different.

The exact concentration-measurement uncertainty for a perfectly monitoring sphere is obtained from the time correlations of molecules inside

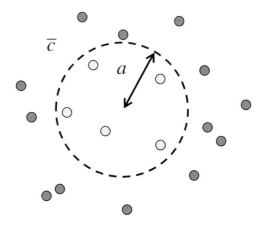

**Fig. 8.4** The perfectly monitoring sphere (model 2) has a ligand-permeable cell surface, and estimates the external ligand concentration from the number of molecules inside its cell volume. Redrawn from Ref. (4).

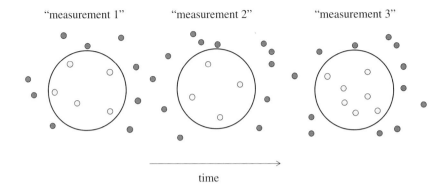

"measurement 1"        "measurement 2"        "measurement 3"

time

**Fig. 8.5** The perfectly monitoring sphere can make multiple "measurements" for increasing its accuracy as long as the measurements are statistically independent.

the sphere[3]. Specifically, we introduce the function $m(t)$, which is the measurement device's output and provides the exact number $l$ of ligand molecules inside the cell volume. This is all the information available to the cell, which has no knowledge about the events outside its volume. If the cell would aim to make an estimate in a time $\tau$, then by time-averaging $l(t)$, i.e.

$$l_\tau = \frac{1}{\tau} \int l(t) dt, \tag{8.4}$$

it would estimate the ligand concentration to $\langle c \rangle = \langle l_\tau \rangle / V$ with the volume given by $(4/3)\pi a^3$. However, if repeated many times, how large is the uncertainty of this estimate? To answer this question we write the variance in ligand-concentration estimation in terms of the variance in counting via

$$\langle (\delta c)^2 \rangle_\tau = \left( \frac{3}{4\pi a^3} \right)^2 [\langle l_\tau^2 \rangle - \langle l_\tau \rangle^2], \tag{8.5}$$

where the mean square value is given by

$$\langle l_\tau^2 \rangle = \frac{1}{\tau^2} \int dt \int dt' \langle l(t) l(t') \rangle = \frac{4\pi}{3} a^3 \bar{c} \cdot u(t - t') + \langle l(t) \rangle \tag{8.6}$$

and $\langle (t) \rangle = \bar{c}V$. In Eq. 8.6 we defined the autocorrelation function $u(t - t') = \langle w(t)w(t') \rangle$, which is the probability that a molecule inside the cell at $t'$ is still (or again) inside the cell at later time $t$ (with $w = 1$ if the molecule is inside and $w = 0$ if molecule is outside the sphere). In fact, only the correlation time $\tau_c = (1/a^2) \int_0^\infty dx\, u(x)$ is required, which can be calculated from solving the diffusion equation in three dimensions and exploiting the analogy with electrostatics.

When all the dust settles, the uncertainty in sensing the ligand concentration for the perfectly monitoring sphere is given by

$$\frac{\langle (\delta c)^2 \rangle_\tau}{\bar{c}^2} = \frac{3}{5\pi Da\bar{c}\tau}, \tag{8.7}$$

which is identical to the estimate in Eq. (8.3) up to a numerical prefactor. Notice, however, that the concentration-measurement uncertainty

[3]For full derivation, see Ref. (2).

of the perfectly absorbing sphere is actually *smaller* than that of a perfectly monitoring sphere of the same size, because the perfectly absorbing sphere removes molecules from the environment and hence does not measure the same molecule more than once.

The perfectly absorbing sphere and the perfectly monitoring sphere appear highly idealized approximations for a real cell. How can they possibly be implemented in a biological context? Cells can become absorbers in multiple ways. They can take up ligand molecules as a way of eating them, they can degrade ligand molecules on their surfaces by secreted enzymes, and more specifically, they can internalize ligand-bound receptors. The latter process is known as endocytosis, but occurs only in cells of higher organisms, not in bacteria. An example for the perfectly monitoring sphere is potentially the well-characterized pathway which pumps a sugar (lactose) into the cell for consumption. Once inside the cell, lactose is effectively measured by the lac repressor, which, when lactose bound, falls off the DNA to allow the expression of key enzymes. The lac repressor hence acts as a cell-internal receptor for the sugar.

How are the above results modified, when considering actual receptors, which can bind and unbind ligand molecules? For simplicity we disregard downstream signaling in the following, but our predictions will nevertheless be rather general.

## 8.3   Sensing with cell-surface receptors

### 8.3.1   Receptor as a counting device

Before discussing sensing by realistic receptors, we introduce the receptor as a kind of counting device for ligand molecules. (Specifically, this will allow us to make a smooth transition from the perfectly monitoring sphere to a realistic receptor.)

Imagine a receptor which can count molecules in its surrounding volume, and from this count can estimate the ligand concentration $c$. Specifically, a molecule is considered "counted" if it was bound by the receptor with the rates for binding and unbinding given by $k_+ \bar{c}$ and $k_-$, respectively. We interpret this reaction as converting "new" (uncounted) molecules into "old" (counted) molecules. Molecules can also enter and leave the volume $V \sim s^3$ via diffusion with rate $k_D \approx D/(2s^2)$ (from now on we drop the factor 2 for back-of-an-envelope calculations). Once an old molecule leaves the volume it is automatically lost into the bulk concentration and can only enter the volume again as a new particle. All these processes are summarized in Fig. 8.6.

Consider the number of old molecules described by the continuous variable $L$. The rate of change of $L$ is then given by

$$\frac{dL}{dt} = k_- n - k_D L - k_+(1-n)L/V, \qquad (8.8)$$

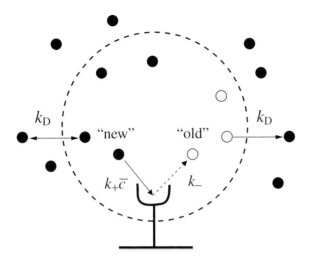

**Fig. 8.6** Introducing the receptor as a counting device. Ligand molecules can enter and leave volume (dashed circle) around receptor by diffusion (rate constant $k_D$) where they are, upon binding, converted from *new* (uncounted, dark gray circles) to *old* (counted, light gray circles) molecules. As indicated the rate of binding is $k_+\bar{c}$ and the rate of unbinding is $k_-$. The receptor estimates the ligand concentration from the number of old molecules in the volume.

where $n$ is the receptor occupancy described by another continuous variable (between 0 to 1). In Eq. (8.8) the first term describes the rate of production of old particles by unbinding, the removal of old particles by diffusion, and the third term the removal of old particles by rebinding. The volume factor is included for unit reasons.

At steady state, Eq. (8.8) can be solved for the old molecules

$$\bar{L} = \frac{k_-\bar{n}}{k_D + k_+(1 - \bar{n})/V} \qquad (8.9)$$

in dependence of the steady-state receptor occupancy $\bar{n}$. We now make the assumption that the uncertainty in ligand concentration $\delta c/\bar{c}$ is given by the Poisson distributed number of old molecules in volume $V$. This leads to

$$\frac{\delta c^2}{\bar{c}^2} = \frac{1}{N_m\bar{L}} = \frac{1}{k_+(1 - \bar{n})\bar{c}\tau} + \frac{1}{Ds\bar{c}\tau}, \qquad (8.10)$$

where the number of measurements $N_m$ conducted in time $\tau$ is given by $k_D\tau$. Eq. (8.10) shows that the uncertainty has two contributions. The first is due to the binding kinetics, describing the rate of conversion from new to old molecules, while the second is due to diffusion, which ultimately limits how fast the receptor receives new molecules for counting.

## 8.3.2 Single realistic receptor

Having gained some intuition about what receptor sensing means, we can now consider a single realistic receptor shown in Fig. 8.7A. The kinetics for the occupancy $n(t)$ of the receptor are given by

$$\frac{dn}{dt} = k_+\bar{c}(1 - n) - k_-n, \qquad (8.11)$$

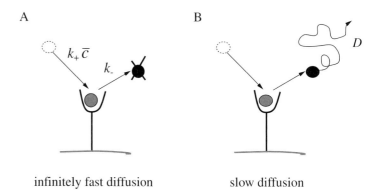

Fig. 8.7 A single receptor which, when unoccupied, can bind a ligand molecule, and, when occupied, can unbind the ligand molecule with rates as indicated. (A) For very fast diffusion a ligand molecule is removed immediately after unbinding, and hence can bind only once. (B) For slow ligand diffusion, the same ligand molecule may bind more than once to the receptor.

where the first term describes ligand–receptor binding and the second unbinding. The steady-state solution for the receptor occupancy is given by

$$\bar{n} = \frac{\bar{c}}{\bar{c} + K_D} \tag{8.12}$$

with $K_D = k_-/k_+$ the ligand dissociation constant. The rates of binding and unbinding are related to the (negative) energy of binding $F$ through detailed balance

$$\frac{k_+\bar{c}}{k_-} = e^{\frac{F}{k_B T}} \tag{8.13}$$

with $k_B T$ the thermal energy. In the limit of very fast ligand diffusion, i.e. when a ligand molecule is immediately removed from the receptor after unbinding, the dynamics of the receptor occupancy is effectively decoupled from the diffusion of ligand molecules, and hence diffusion does not need to be included explicitly.

The accuracy of sensing is obtained by applying the *fluctuation–dissipation theorem* (FDT) from Chapter 3 (illustrated in Fig. 8.8). The FDT relates the spectrum of the fluctuations in occupancy $S_n(\omega) = \langle \delta\hat{n}(\omega)\delta\hat{n}^*(\omega)\rangle$ at thermal equilibrium to the linear response $\Delta n$ to a perturbation in the receptor binding energy $\Delta F$ via

$$S_n(\omega) = \frac{2k_B T}{\omega} \text{Im}\left[\frac{\Delta\hat{n}}{\Delta\hat{F}}\right], \tag{8.14}$$

Fig. 8.8 Application of the fluctuation–dissipation theorem (FDT) to the accuracy of sensing by a single receptor. The uncertainty of ligand concentration $\delta c$ can be inferred from the uncertainty in ligand-receptor occupancy $\delta n$ by calculating the deterministic linear response of the occupancy $\Delta n$ to a perturbation of the ligand-receptor energy $\Delta F$. Approach based on Ref. (3).

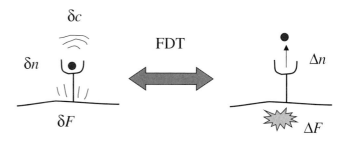

where Im indicates imaginary part and the ^ symbol indicates the Fourier transform from the time to the frequency ($\omega$) domain. The term in brackets on the right-hand side of Eq. (8.14) is the generalized susceptibility, and its imaginary part determines how much energy is dissipated as heat. Furthermore, at equilibrium the uncertainty in ligand concentration can be directly related to the fluctuations in occupancy using Eq. (8.12)

$$\delta c = \frac{k_-}{k_+(1-\bar{n})^2}\delta n. \tag{8.15}$$

For a measurement performed on a time scale $\tau$ much larger than the correlation time of the binding and unbinding events, $\tau_c = (k_+\bar{c}+k_-)^{-1}$, the fluctuations of the occupancy are obtained from the zero-frequency spectrum $S_n(\omega = 0) = 2k_+(1-\bar{n})\bar{c}\tau_c^2$ divided by $\tau$. Putting the pieces together,[4] we finally obtain the uncertainty in measuring ligand concentration

$$\frac{\langle(\delta c)^2\rangle_\tau}{\bar{c}^2} = \frac{2}{k_+\bar{c}(1-\bar{n})\tau} \to \frac{1}{2\pi D\bar{c}s\tau}, \tag{8.16}$$

[4]For full derivation, see Ref. (3).

where the right-hand side is obtained for diffusion-limited binding, i.e. when $k_+\bar{c}(1-\bar{n}) \to 4\pi\bar{c}Ds$, with $D$ the ligand diffusion constant and $s$ the dimension of the (spherical) receptor. Eq. (8.16) shows that the accuracy of sensing is limited by the random binding and unbinding of ligand molecules.

When diffusion of ligand is slow, ligand binding to the receptor is affected by diffusion, since ligand molecules are not immediately removed after unbinding. This scenario is indicated in Fig. 8.7B. The kinetics of the receptor occupancy and ligand concentration are then described by

$$\frac{dn(t)}{dt} = k_+c(\vec{x}_0,t)[1-n(t)] - k_-n(t) \tag{8.17}$$

$$\frac{\partial c(\vec{x},t)}{\partial t} = D\nabla^2 c(\vec{x},t) - \delta(\vec{x}-\vec{x}_0)\frac{dn(t)}{dt}, \tag{8.18}$$

where we explicitly kept the time and space dependence of the variables for increased clarity. Furthermore, $\vec{x}_0$ indicates the position of the receptor, and $\delta(\vec{x}-\vec{x}_0)$ is the Dirac $\delta$-function centered at the receptor location. The last term in the second equation describes a sink or source of ligand at $\vec{x}_0$, corresponding to ligand-receptor binding or unbinding, respectively. Analogous to fast diffusion, Eqs. (8.17) and (8.18) have the respective steady-state solutions $\bar{n}$ given by Eq. (8.12) and $\bar{c}$ independent of $D$.

Following a slightly more elaborate (but otherwise very similar) procedure as in the previous case with fast diffusion, the uncertainty of sensing is given by

$$\frac{\langle(\delta c)^2\rangle_\tau}{\bar{c}^2} = \frac{2}{k_+\bar{c}(1-\bar{n})\tau} + \frac{1}{\pi sD\bar{c}\tau} \tag{8.19}$$

$$\to \frac{3}{2\pi sD\bar{c}\tau}, \tag{8.20}$$

where the first term on the right-hand side of Eq. (8.19) is the same as in Eq. (8.16), while the second term is the increase in uncertainty due to diffusion. This term accounts for the additional measurement uncertainty from rebinding of previously bound ligand to the receptor. (Note that Eq. (8.19) is very similar—up to factors of 2 and $\pi$—to our previous heuristic estimate from Eq. (8.10).) For diffusion-limited binding, we obtain Eq. (8.20).

Comparison of Eqs. (8.16) and (8.20) shows clearly that removal of previously bound ligand by fast diffusion increases the accuracy of sensing, since the same ligand molecule is never measured more than once. As a result, the calculation with ligand-receptor binding with and without coupling to diffusion mirror the result from the perfectly monitoring and perfectly absorbing spheres. The limit of diffusion-limited binding makes it clear that the accuracy is ultimately limited by the random arrival of ligand molecules at the receptor by diffusion, irrespective of details of sensing and signaling.

### 8.3.3   Multiple realistic receptors

The single-receptor model can be extended to multiple receptors as illustrated in Fig. 8.9. Here we consider $N_T$ evenly distributed receptors on a sphere for which an analytical result can be derived. Following a very similar procedure to the single receptor with slow diffusion, we obtain[5]

$$\frac{\langle(\delta c)^2\rangle_\tau}{\bar{c}^2} = \frac{2}{N_T k_+ \bar{c}(1-\bar{n})\tau} + \frac{1}{N_T \pi D \bar{c}\tau}\left(\frac{\pi}{s} + \frac{\phi}{2}\right). \tag{8.21}$$

When compared to the single receptor, the right-hand side of Eq. (8.21) depends inversely on the number of receptors, $N_T^{-1}$, which describes spatial averaging. Furthermore, the second term has two contributions. The first describes rebinding to the same receptor known from before, while the second describes rebinding to any of the other receptors via structure factor $\phi$. This parameter includes information on the spatial distribution of the receptors. Due to this last contribution to rebinding,

[5]For full derivation see Ref. (3).

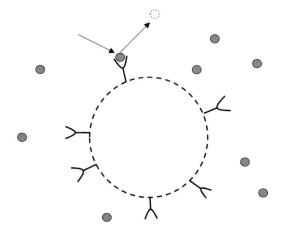

**Fig. 8.9** A slightly more realistic cell (model 3) estimates the external ligand concentration from the occupancies of its cell-surface receptors.

the uncertainty of sensing depends on the spatial distribution of the receptors—the closer they are together, the larger this noise.

These results show that for a cell the accuracy of sensing (the inverse of the calculated uncertainty) is quite different from the types of measurements we are used to from our macroscopic world. For instance, if we measure the length of an object, e.g. of a pencil, then the uncertainty is often determined by systematic errors in the ruler itself or by subjective errors, e.g. how well we are able align the straight ruler with the 3D object to be measured (problem of slanted tip of the pencil). In the microscopic world of the cell instead, individual molecules are measured (effectively counted) and from this noisy measurement the average ligand concentration is inferred. This type of Poisson-like noise depends on the background concentration, making it more and more difficult to sense lower and lower concentrations in an available time interval. This time interval is often extremely short, of order of a second or less, and can determine if the cell finds new sources of food or is able to escape a toxin.

# Chapter summary

- What limits the accuracy by which a cell can sense chemicals in its environment? While there are cell-internal sources of noise, the accuracy is ultimately limited by cell-external noise from the random arrival of ligand molecules at the receptors by diffusion. To quantitatively predict the fundamental physical limit on the detection of ligand molecules, i.e. the accuracy of sensing, we considered three different models.

- **Model 1: perfectly absorbing sphere**: The perfectly absorbing sphere of size $a$ has a concentration-measurement uncertainty of

$$\frac{\langle (\delta c)^2 \rangle_\tau}{\bar{c}^2} = \frac{1}{4\pi D a \bar{c} \tau} \tag{8.22}$$

with $D$ the ligand diffusion constant and $\tau$ the averaging time. Hence, the accuracy can be improved by increased time averaging, faster diffusion, as well as more spatial averaging with an increased measurement device size.

- **Model 2: perfectly monitoring sphere**: The uncertainty in ligand concentration for the perfect monitor is given by

$$\frac{\langle (\delta c)^2 \rangle_\tau}{\bar{c}^2} = \frac{3}{5\pi D a \bar{c} \tau}. \tag{8.23}$$

This expression depends again on the same physical parameters, but due to its prefactor the perfect monitor has a larger uncertainty than the perfectly absorber.

- **Model 3: sensing with cell-surface receptors**: A single receptor of dimension $s$ with ligand binding and unbinding rate constants $k_+$ and $k_-$ has an uncertainty

$$\frac{\langle (\delta c)^2 \rangle_\tau}{\bar{c}^2} = \frac{2}{k_+ \bar{c}(1 - \bar{n})\tau} + \frac{1}{\pi s D \bar{c} \tau}, \qquad (8.24)$$

  where the second term is due to rebinding when diffusion is sufficiently slow.

- Based on these results, ligand-absorbing cells sense most accurately. Cells could technically achieve this by ligand-receptor internalization, uptake (eating), or ligand degradation on cell surface by enzymes. Our results are not only important for bacteria to find food or to escape a toxin. The accuracy of sensing is also a fundamental parameter determining the success of our immune system, embryonic development, and functioning of the brain.

# Further reading

(1) *Random Walk in Biology*, by Howard C. Berg (Princeton University Press, 1993).

(2) 'Physics of chemoreception', by Howard C. Berg and Edward M. Purcell, *Biophys. J.* 20: 193–219 (1977).

(3) 'Physical limits to biochemical signaling' by William Bialek and Sima Setayeshgar, *Proc. Natl. Acad. Sci. USA* 102: 10040–45 (2005).

(4) 'Accuracy of direct gradient sensing by single cells' by Robert G. Endres and Ned S. Wingreen, *Proc. Natl. Acad. Sci. USA* 105: 15749–54 (2008).

(5) 'Increased accuracy of ligand sensing by receptor diffusion on cell surface', by Gerardo Aquino and Robert G. Endres, *Phys Rev E* 82: 041902 (2010).

(6) 'Fundamental limits on the suppression of molecular fluctuations' by Ioannis Lestas, Glenn Vinnicombe, and Johan Paulsson, *Nature* 467: 174–8 (2010).

# Motor impulse response

9

Biological cells have to deal with computational challenges similar to the ones we encounter in electrical engineering or telecommunication. Both biological and engineered systems need to transmit signals reliably, so an appropriate response can be launched by the cell or the end user. In particular, signals must be prevented from corruption by noise. However, if we interpret the biological pathway as a signal processing network, we need to keep in mind that components in the pathway are not hard-wired but are diffusive, and that reactions are based on random chemical events. In analogy to engineering, we can ask what types of stimuli are best transmitted through the pathway and how noise is filtered out. Additionally, we might want to know how the stimulus is modified by the cell and its signal transduction pathway, and how the evolutionary pathway design reflects the tasks the cell tries to perform.

## 9.1 Impulse response

System components in telecommunication or parts of a computer chip are generally well characterized individually before they are used in a device to solve a particular task. However, such a bottom-up approach for understanding a biological system may not always be feasible, e.g. back in time when the experimental characterization of the *E. coli* chemotaxis proteins was not possible yet. Under these circumstances, a black-box approach may be applied as illustrated in Fig. 9.1. In this top-down approach the system is perturbed, and the response behavior is characterized without the need to know all the components in detail. Once the global behavior of the system is known, a more detailed approach can fill in the gaps later.

To conduct such an investigation of the chemotaxis pathway, a powerful conceptual framework to use is *linear impulse response theory* (Chapter 3). In this framework a sharp (in time) external stimulus is applied and the final cellular output is measured. (Of course, more can be learned if intermediate steps can be measured, e.g. by FRET.) The impulse response has two attractive attributes, making it a very useful tool. First, only the time constants of the processes occurring in the system matter, not the time constants of the stimulus itself since the pulse is so short. Second, the impulse response contains all the information needed to predict the response to more complicated time-varying stimuli, since we can always decompose a stimulus into a sequence of short pulses, provided the amplitudes are adjusted appropriately. However, for

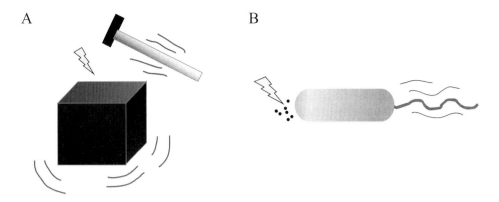

**Fig. 9.1** Illustration of the black-box approach to characterize the response of a system irrespective of the underlying details. Shown are both a macroscopic (A) and a microscopic cellular (B) example.

**Fig. 9.2** Experimental generation of sharp ligand pulse. (A) A short release of attractant chemical is repelled in the electric field from pipette, thus creating a density wave of ligand at each cell location. (B) Concentration of ligand as seen by an individual cell in time. Pulse is on top of a background concentration. Experiment based on Ref. (1).

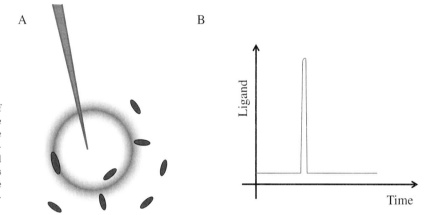

this assembly approach to work, the stimulus must be small enough so that there are no non-linear effects, which would mask the output characteristics. Furthermore, the response to each impulse must be independent of the previous impulse.

In *E. coli* chemotaxis, a short ligand pulse can be generated by an iontophoretic technique: Chemoattractant is briefly released and repelled from an electrically charged pipette. This creates a diffusive wave of ligand molecules as illustrated in Fig. 9.2. The final motor output can be measured by tethering individual flagella to a microscope glass slide, allowing the experimentalist to observe the sequence of clockwise (CW) and counterclockwise (CCW) rotations (see Fig. 9.3). The CW and CCW time intervals are approximately exponentially distributed, which indicates that the motor switching process is close to a memory-less two-state Poisson process. Averaging over many binary time traces is necessary to obtain a smooth curve of the motor response for easier interpretation. Difficulties can arise when the cell-to-cell variation is very large, and the length of the time traces is limited by the time scale of gene expression or cell division.

A  Tethered-cell experiment

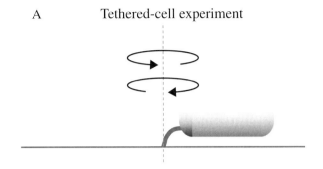

B

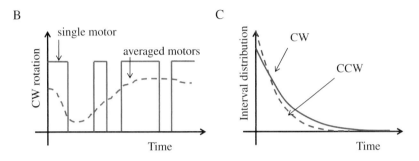

C

**Fig. 9.3** Tethered-cell assay. (A) Rotations of an individual motor. (B) Time trace of a single motor and averaged time traces over many motors. (C) The interval distribution is approximately exponential. CW and CCW stand for clockwise and counterclockwise motor rotation. Experiment based on Ref. (1).

## 9.2 Time and frequency domains

The results of the impulse-response experiment of tethered cells are sketched in Fig. 9.4. In panel A, the average motor CW bias in response to a short impulse at time 0s is shown. Note that CW rotation corresponds to the cellular "tumble" mode if cells were freely swimming. Just by inspection, several features of the chemotaxis pathway can be deduced. The impulse response has a substantial width of about 4s, indicating that cells integrate stimuli for a few seconds in determining the bias. Consequently, cells must have a short-term memory (which of course is provided by receptor methylation). The impulse response is also biphasic, with one lobe below the baseline and the other above it. This shows that cells can make differential measurements, that is they can differentiate an increase in ligand concentration from a decrease. If the areas under the two lobes are equal, the pathway is fully adaptive and precise (Chapter 6). A bit harder to see is that the response occurs about a fraction of a second after the actual impulse. This response latency is longer than the time it takes for a molecule to diffuse the distance from the receptors to the motors (about 1 $\mu$m, and hence is not modeled explicitly). This indicates the time required to process the signal along the pathway.

Fig. 9.4B shows a Bode plot, which is a double logarithmic plot of the magnitude of the Fourier transform (the conversion of the time-dependent response to the frequency domain). Such a plot shows the frequency components of the response if it were periodic in time (see Appendix A). By plotting such a spectrum, additional insights into the

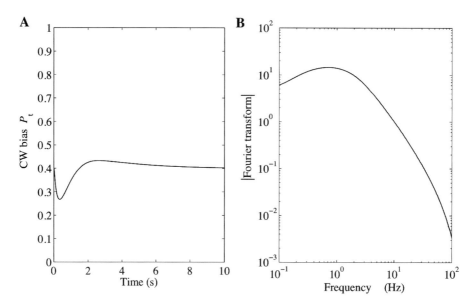

**Fig. 9.4** Motor impulse response. (A) Response of motor CW tumble bias to a sharp ligand pulse at time 0s. (B) Modulus (magnitude) of the corresponding Fourier transform (Bode plot). There is a good match between these calculated curves and actual experimental data from Ref. (1).

[1] This analysis is based on Ref. (1).

pathway can be gained.[1] First, the spectrum has bandpass properties with a well-defined maximum, corresponding to the inverse of the integration time of a few seconds. This shows that the pathway is maximally sensitive to changes occurring with this periodicity and reflects the gradients the cells typically experience when swimming with an average run time of about 1s. Furthermore, the pathway can be described in terms of filters as done in electrical engineering by considering the slopes of the spectrum. A positive slope indicates a high-pass filter, a "device" which only allows high-frequency stimuli to pass through. In contrast, a negative slope indicates a low-pass filter, which allows only low-frequency stimuli to pass through the pathway. Looking more carefully at the slopes of actual data shows that the chemotaxis pathway has first-order high-pass and third-order low-pass filter characteristics. The three filters at high frequencies should reflect the details of the pathway to be investigated next. Both types of filters put together result in band-pass characteristics: that is, the pathway transmits signals over a certain range of frequencies.

## 9.3   Minimal pathway model

How can these observations be understood theoretically? Here we use a minimal model, which can be solved analytically while still providing valuable insight into signal propagation (see Fig. 9.5). Specifically, we describe receptor signaling by the receptor-complex activity from Chapter 5,

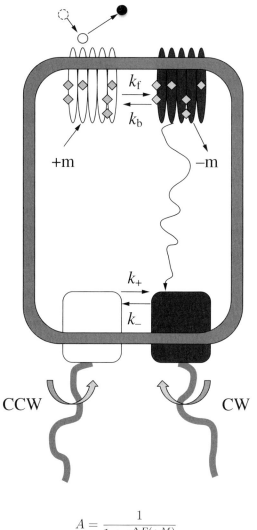

**Fig. 9.5** Minimal model of the chemotaxis pathway in *E. coli*, including sensing of ligand concentration by receptor complexes, adaptation by receptor methylation and demethylation, and regulation of the motor. Specific emphasis is put on the noise sources in the pathway, including extrinsic ligand noise and intrinsic noises from random receptor-complex switching, methylation/demethylation events, and motor switching. Redrawn from Ref. (4).

$$A = \frac{1}{1 + e^{\Delta F(c,M)}} \qquad (9.1)$$

with $c$ the ligand concentration and $M$ total receptor-complex methylation level. Furthermore, the dynamics of the receptor-complex methylation level were given in Chapter 6. Here we use a simplified version of Eq. (6.9), namely

$$\frac{dM}{dt} = N_R g_R (1 - A) - N_B g_B A, \qquad (9.2)$$

assuming a linear dependence of the rates in $A$.

In this minimal model we "neglect" the fast phosphorylation and dephosphorylation reactions. Specifically, the degree of phosphorylation of proteins CheA, CheB, and CheY are approximately proportional to the receptor activity $A$ assuming quasi-steady state. However, we now require the switching (that is the rate constants) to depend on the activity directly.

The switching dynamics for the fraction of CW rotating motors, $P_{CW}$, is given by

$$\frac{dP_{CW}}{dt} = k_+(A) \cdot (1 - P_{CW}) - k_-(A) \cdot P_{CW} \qquad (9.3)$$

with the $k_+$ and $k_-$ the activity-dependent rate constants for switching from CCW to CW and from CW to CCW, respectively. For constant activity $A$, Eq. (9.3) describes a simple Poisson process, which leads to exponential distributions for the CW and CCW interval lengths (see Fig. 9.3C).

However, so far the dynamics of the ligand concentration are still missing. To describe a sharp pulse in ligand concentration, we use

$$c(t) = \bar{c} + \Delta c_t \delta(t - t_0) \qquad (9.4)$$

with $\bar{c}$ describing the background concentration and the Dirac $\delta$-function describing the pulse at time $t = t_0$ with integrated ligand concentration $\Delta c_t$. To describe how the methylation and ligand dynamics translate into the dynamics of the activity, we use

$$\frac{dA}{dt} = \frac{\partial A}{\partial M}\frac{dM}{dt} + \frac{\partial A}{\partial c}\frac{dc}{dt}, \qquad (9.5)$$

since the corresponding changes in activity are nearly instantaneous. With this model we can study signal propagation and noise filtering in the chemotaxis pathway.

## 9.4   Linear response approximation

In linear response theory (see Chapters 3 and 6), we can write for the response of a quantity of interest in the pathway

$$R(t) = \int_{-\infty}^{t} \chi(t - t')c(t')dt', \qquad (9.6)$$

where $\chi(t - t')$ is the linear response function and $c(t')$ is the perturbation in ligand concentration. After Fourier transformation, the integral deconvolutes and we obtain the simple product

$$\hat{R}(\omega) = \hat{\chi}(\omega)\hat{c}(\omega), \qquad (9.7)$$

allowing us to solve for the response function

$$\hat{\chi}(\omega) = \frac{\hat{R}(\omega)}{\hat{c}(\omega)}. \qquad (9.8)$$

The Fourier-transformed response function is generally complex with a real and an imaginary part (see Appendix A). These can also be written in terms of the magnitude $|\hat{R}(\omega)|$ and phase $\phi(\omega)$.

Specifically, the above equations for the change in activity can be solved by linearization around the steady-state value $A(t) = \bar{A} + \Delta A(t)$

and Fourier transformation. Plugging Eq. (9.2) into Eq. (9.5), followed by linearization, leads to

$$\frac{d(\Delta A)}{dt} = -\underbrace{(N_R g_R + N_B g_B)\frac{\partial A}{\partial M}}_{=\lambda_M}\Delta A + \frac{\partial A}{\partial c}\frac{dc}{dt}, \qquad (9.9)$$

where we introduced the relaxation rate $\lambda_M$ due to adaptation. To get rid of the time derivatives, we Fourier-transform and obtain for the magnitude of the response

$$\frac{|\Delta \hat{A}|}{\Delta c_t} = \frac{\omega\frac{\partial A}{\partial c}}{\sqrt{\lambda_M^2 + \omega^2}}, \qquad (9.10)$$

which we make independent of the stimulus strength through normalization. This shows that the receptor activity is a high-pass filter—the magnitude of the response becomes increasingly small for frequencies below the characteristic frequency of adaptation, $\lambda_M$. However, above this frequency the response is constant, determined by the sensitivity of the receptors to ligand.

To describe the response of the motor, we need to linearize Eq. (9.3) for the motor dynamics. In addition to $P_{CW}(t) = \bar{P}_{CW} + \Delta P_{CW}(t)$, we also need to linearize the switching rates around the steady-state activity $k_+(t) = \bar{k}_+ + \Delta k_+(t) = \bar{k}_+ + \frac{\partial k_+}{\partial A}\Delta A(t)$ and similarly for $k_-$.[2] Hence, starting from Eq. (9.3), we finally obtain for the (normalized) motor response

[2]This technically requires a description of how the rates depend on the activity, which we omit here. For details see Ref. (4).

$$\frac{|\Delta \hat{P}_{CW}|}{\Delta c_t} = \frac{g_p|\Delta\hat{A}|/\Delta c_t}{\sqrt{\omega^2 + \kappa_{CW}^2}} \qquad (9.11)$$

$$= \frac{\omega g_p(\partial A/\partial c)}{\sqrt{(\omega^2 + \lambda_M^2)(\omega^2 + \kappa_{CW}^2)}} \qquad (9.12)$$

with the characteristic frequency of the motor $\kappa_{CW} = \bar{k}_+ + \bar{k}_-$ and motor sensitivity $g_p = (1 - \bar{P}_{CW})(dk_+/dA) - \bar{P}_{CW}(dk_-/dA)$. Eq. (9.11) shows that the motor by itself is a low-pass filter, that is it produces a constant response for frequencies below $\kappa_{CW}$ but filters out signals above this frequency. Together with the upstream high-pass filter, the final cell output, Eq. (9.12), is a band-path filter.

The propagation of the response down the pathway to the motor is shown in Fig. 9.6 (middle panels). When considering the magnitude of the response, we observe the filtering of low-frequency contributions due to adaptation, which effectively removes slowly varying signals in time and prevents cells from responding. In contrast, the filtering of high-frequency contributions is the result of the finite rates of the individual reactions in the pathway, which limits the temporal resolution of the pathway for responding to rapidly changing signals.

Looking at the phase of the Fourier transform is interesting as well. As shown in the same figure, the phase is close to $\pi/2$ at low frequencies for receptor-complex activity and further down the pathway. This is

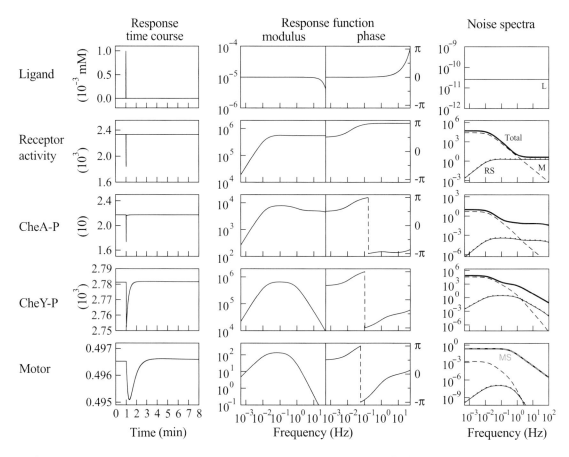

**Fig. 9.6** Calculated transmission of ligand pulse along chemotaxis pathway in time (left) and frequency (modulus and phase; middle) domains. (right) Noise power spectra along pathway. Calculations are done with full pathway model, which, in addition to the reactions of the minimal model, also includes phosphorylation of CheA, CheB and CheY. Shown are contributions from ligand (L), methylation/demethylation (M), receptor-complex switching (RS) and motor switching (MS). Redrawn from Ref. (4).

indicative of adaptation, as this allows the receptors to effectively take the time derivative of the stimulus. In contrast for high frequencies, the phase shift tends towards $\pi$ for the receptor activity. This is a result of fact that receptors in this limit simply follow the stimulus with an opposite sign (activity drops for increase in ligand). A bit for CheA-P, and more for CheY-P and the motor, the phase goes significantly beyond $\pi$, indicating a lag due to the slower response in these pathway components.

## 9.5   Noise power spectra

Another way of studying the pathway is by doing nothing, or in other words, by investigating the noise of the unstimulated motor. From long time traces of the binary motor switching, we can calculate the auto-correlation function

$$K(\tau) = \langle \delta P_{\mathrm{CW}}(t) \delta P_{\mathrm{CW}}(t + \tau) \rangle, \tag{9.13}$$

which is the average value of the product of fluctuations at two time points. This function only depends on the interval length between time points if the dynamics of $P_{\mathrm{CW}}(t)$ are stationary.

Using the Wiener–Kinchin theorem, the power spectrum is the Fourier transform of the auto-correlation function

$$S_{\mathrm{CW}}(\omega) = \int_{-\infty}^{\infty} K(t) e^{i\omega t} dt \tag{9.14}$$

(see Chapter 3 and Appendix A). For exponentially decaying correlations, such as the motor interval distributions (waiting times) in Fig. 9.3, a Lorentzian spectrum of the form $S(\omega) \sim 1/(\omega^2 + q^2)$ with cut-off frequency $q$ is expected. However, as signaling noise upstream of the motor may affect the motor switching, deviations from a simple Lorentzian are expected to occur. This was indeed seen in experiments (see Fig. 9.8A below). The spectrum is Lorentizan when a constitutively active CheY mutant is used, which is not regulated by kinase CheA and hence does not exhibit fluctuations in phosphorylation level. However, when the signaling pathway is intact, a low-frequency peak is observed, here due to slow methylation and demethylation events.

To investigate the noise power spectrum of the motor, we need to include noise in our simplified model of the signaling pathway. Mathematically, this is most easily achieved by adding Langevin noise terms to the deterministic differential equations. After linearization of the equations around their steady-state values and subsequent Fourier transformation, the power spectra are obtained (see Chapter 4). Specifically, we linearize by introducing the following fluctuations: $A(t) = \bar{A} + \delta A(t)$ for the activity, $\delta c = \bar{c} + \delta c$ for the ligand concentration, $k_+ = \bar{k}_+ + \delta k_+(A)$ and $k_- = \bar{k}_- + \delta k_-(A)$ for the activity-dependent motor switching rates, and $P_{CW} = \bar{P}_{CW} + \delta P_{CW}$ for the motor bias. The resulting linearized equations are

$$\frac{d(\delta A)}{dt} = \frac{\partial A}{\partial M} \frac{d(\delta M)}{dt} + \frac{\partial A}{\partial c} \frac{d(\delta c)}{dt} + \eta_a \tag{9.15}$$

for the receptor-complex activity with noise term $\eta_a$ from random on/off switching,

$$\frac{d(\delta M)}{dt} = -(N_R g_R + N_B g_B)\delta A + \eta_M = \lambda_M \delta M + \eta_M \tag{9.16}$$

for the methylation dynamics using noise term $\eta_M$ from random methylation and demethylation events,

$$\frac{d(\delta c)}{dt} = \eta_c \tag{9.17}$$

for the ligand dynamics with noise term $\eta_c$ to describe the ligand noise, and

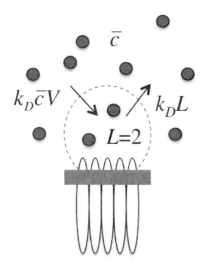

$$\frac{d(\delta P_{\text{CW}})}{dt} = \left[(1 - \bar{P}_{\text{CW}})\frac{dk_+}{dA} + \bar{P}_{\text{CW}}\frac{dk_-}{dA}\right]\delta A - (\bar{k}_+ + \bar{k}_-)\delta P_{\text{CW}} + \eta_{\text{CW}}$$

(9.18)

for the dynamics of motor CW/CCW switching with noise term $\eta_{\text{CW}}$. Note that in the minimal model Eq. (9.18) depends directly on the upstream activity.

After Fourier transformation, the power spectra, e.g. of the motor, is obtainable from

$$S_{\text{CW}}(\omega) = \langle \delta\hat{P}_{\text{CW}}(\omega)\delta\hat{P}_{\text{CW}}^*(\omega)\rangle,$$

(9.19)

which, up to time units, is identical to the previous definition in Eq. (9.19).

To show how the calculation of the power spectra is done in practice, we start with the activity of the receptor complex and its noise sources. To describe the ligand noise we technically have to consider the random binding and unbinding of ligand molecules to two receptor-activity states (see Chapter 8 for the treatment of a one-state receptor). However, to simplify this problem, while still capturing the essence behind this noise, we consider simple fluctuations in ligand concentration (density) in the vicinity of the receptors. Fig. 9.7 shows ligand molecules entering volume $V \approx s^3$ of dimension $s$ with rate $k_D\bar{c}V$ and leaving the volume with rate $k_D L$, where $k_D \approx D/s^2$ is the rate of diffusion. This type of noise describes the fluctuations in ligand concentration due to diffusion and can be considered more fundamental than the ligand-receptor binding and unbinding noise. Hence, we can write for the number $L$ of ligand molecules in this volume

$$\frac{dL}{dt} = k_D(\bar{c}s^3 - L) + \eta_L(t),$$

(9.20)

where $\eta_L$ is a Gaussian white noise with zero mean and noise intensity $Q_L = 2Ds\bar{c}$ (see Chapter 4). Linearization and Fourier transformation leads to

$$S_L(\omega) = \frac{2Ds\bar{c}}{\omega^2 + k_D^2}$$

(9.21)

and $S_c(\omega) = S_L(\omega)/V^2$ in terms of ligand concentration. Hence, in the zero-frequency limit we recover formula $\langle(\delta c)^2\rangle_\tau/\bar{c}^2 \approx 2/(D\bar{c}s\tau)$, which, up to a prefactor, was provided in Chapter 8 for the time-averaged (by time $\tau$) ligand noise in the diffusion-limited case.

To calculate the receptor-complex switching noise, we write a dynamical equation for the receptor-complex activity $a$ (similar to what we previously did for the motor switching)

$$\frac{da}{dt} = k_f(1 - a) - k_b a + \eta_a$$

(9.22)

with $\bar{a} = \frac{k_f}{k_f + k_b}$ identical to $\bar{A}$, and apply one more time the Langenvin approximation. This leads to the noise power spectrum for the receptor-complex switching

**Fig. 9.7** Illustration of ligand noise at a receptor complex. The rate of ligand arrival by diffusion is given by $k_D\bar{c}V$ with $k_D$ the rate of diffusion, $\bar{c}$ the background ligand concentration, and $V$ a small volume element at the receptor complex (dashed line). Ligand molecules leave the volume with rate $k_D L$, where $L$ is the number of ligand molecules in $V$ (here $L = 2$).

$$S_a(\omega) = \frac{2k_f \bar{A}}{\omega^2 + (k_f + k_b)^2}. \tag{9.23}$$

As a result the ligand and switching power spectra (Eq. (9.21) and Eq. (9.23) respectively) are Lorentzian, and constitute colored[3] input noise for the rest of the signaling pathway.

[3]White noise contains all frequency components equally, while colored noise has strong bias in certain frequency components.

We first calculate the power spectrum of the total activity. Considering $N_C$ receptor complexes in the cell, we obtain

$$S_A(\omega) = N_C \frac{\omega^2 \left[ S_a(\omega) + \left(\frac{\partial A}{\partial c}\right)^2 S_c(\omega) \right] + \left(\frac{\partial A}{\partial M}\right)^2 Q_M}{\lambda_M^2 + \omega^2}. \tag{9.24}$$

In this equation, the first term arises due to receptor-complex switching with $S_a(\omega)$ given by Eq. (9.23), the second term represents the amplified ligand noise from diffusion with $S_c(\omega)$ given by Eq. (9.21) (up to normalization by the volume), and the third term is the methylation and demethylation noise with noise intensity $Q_M = 2N_R \langle |\hat{\eta}_M|^2 \rangle = 2N_R g_R (1 - \bar{A})$ as each tethered enzyme contributes noise. We also assumed that all the $N_C$ receptor complexes in a cell act independently, so their noise contributions add up.

Analogously, using Eq. (9.19) we obtain for the motor noise power spectrum

$$S_{\mathrm{CW}}(\omega) = \frac{g_p^2 S_A(\omega) + Q_{\mathrm{CW}}}{\omega^2 + \kappa_{\mathrm{CW}}^2} \tag{9.25}$$

with $g_p$ and $\kappa_{\mathrm{CW}}$ as defined before. In this equation the second term in the numerator leads to the overall Lorentzian motor spectrum with motor noise intensity $Q_{\mathrm{CW}} = 2k_+(\bar{A})(1 - \bar{P}_{\mathrm{CW}})$. In contrast, the first term represents the amplified (by factor $g_p$ in numerator) and filtered (by frequency $\kappa_{\mathrm{CW}}$ in denominator) upstream noise.

The model makes the prediction that the various noise sources in the pathway have characteristic signatures in the motor spectrum. Fig. 9.6 (right panels) shows that these are generally very small compared to the dominant Lorentzian motor spectrum. However, due to cell-to-cell variation, we would expect the some cells have a less dominant Lorentzian motor spectrum, e.g. due to faster motor switching at a small steady-state bias. The motors of these cells can show visible signatures of methylation and demethylation noise at low frequencies, exemplified by the low-frequency bump in the calculated spectrum of Fig. 9.8A.

Additionally, expression levels of proteins may fluctuate from cell to cell. Specifically, the model predicts that the low-frequency bump depends on the expression level of the adaptation enzymes with overexpression of CheR significantly reducing the bump (Fig. 9.8B). Enhanced noise at low expression levels may allow subpopulations of cells to have longer runs and thus to explore more distant regions in space. As a result, a population of cells can "hedge their bets" with some being more "advantageous" than others.

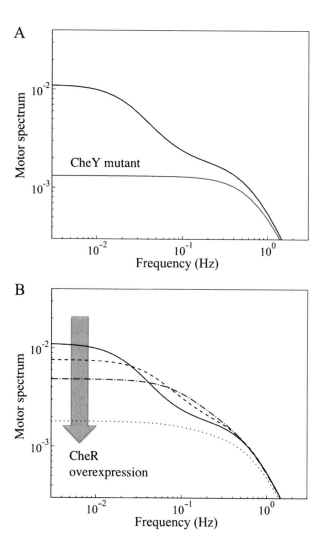

**Fig. 9.8** Variability of motor power spectrum. (A) Spectrum of cells with normal (wild-type) CheY (black) and constitutively active (unregulated) CheY mutant (gray). (B) Motor spectrum for increasing CheR expression levels, including 1x (solid), 2x (dashed), 4x (dashed-dotted), and 10x (dotted) wild-type levels. These calculated spectra are in good agreement with experiments. Redrawn from Ref. (4), and based on data from Ref. (3).

# Chapter summary

- Similar to an engineered device, a cell with its signal transduction pathways can be probed using external stimuli, e.g. chemicals, and by measuring the cell's output or response. Linear response theory is a particularly convenient conceptual framework for analysis and model calculations. Specifically, using a sharp stimulus pulse, the pathway can be probed at all frequencies simultaneously.

- The signal power spectrum of a single motor has a peak at the typical frequency corresponding to the inverse run length between tumbles. It acts as a band-pass filter. Low-frequency signals are filtered out due to adaptation, while high-frequency signals are filtered out due to the finite reaction rates in the pathway, such as phosphorylation reactions. In a simple model of the pathway in which the receptor activity directly regulates the motor switching rate, the response is given by

$$\frac{|\Delta \hat{P}_{CW}|}{\Delta c_t} = \frac{\omega g_A g_P}{\sqrt{(\omega^2 + \lambda_M^2)(\omega^2 + \kappa_{CW}^2)}}, \qquad (9.26)$$

where $g_A$ and $g_P$ are the receptor complex and motor sensitivities, respectively, and $\lambda_M$ and $\kappa_{CW}$ are the cut-off frequencies due to methylation and motor switching, respectively. Together with the $\omega$ factor in the numerator, this gives rise to a band-pass filter.

- The noise power spectrum is dominated by the random motor switching but has small signatures of all upstream reactions in the pathway. The motor switching cannot really be considered noise, but instead is the means of the cell to move and search for food. In the simple model from above, the motor power spectrum is given by

$$S_{CW}(\omega) = \frac{g_P^2 S_A(\omega) + Q_{CW}}{\omega^2 + \kappa_{CW}^2}, \qquad (9.27)$$

where $S_A$ is the power spectrum of the receptor complex and $Q_{CW}$ is the noise strength of the motor. This gives rise to a modified Lorentzian spectrum.

- The experimentally determined noise power spectra, e.g. using the tethered cell assay, shows significant cell-to-cell variation. The variation in motor behavior may lead to increased fitness of the cell population, as some cells may reach more distant higher nutrient concentrations more quickly.

# Further reading

(1) 'Impulse responses in bacterial chemotaxis', by Steven M. Block, Jeffrey E. Segall, and Howard C. Berg, *Cell* 31: 215–26 (1982).

(2) 'Adaptation kinetics in bacterial chemotaxis', by Steven M. Block, Jeffrey E. Segall, and Howard C. Berg, *J. Bacteriol.* 154: 312–23 (1983).

(3) 'From molecular noise to behavioural variability in a single bacterium', by Ekaterina Korobkova, Thierry Emonet, Jose M. G. Vilar, Thomas S. Shimizu, and Philippe Cluzel, *Nature* 428: 574–8 (2004).

(4) 'Noise characteristics of the *Escherichia coli* rotary motor', by Diana Clausznitzer and Robert G. Endres, *BMC Syst. Biol.* 5: 151 (2011).

(5) 'Modeling the chemotactic response of *Escherichia coli* to time-varying stimuli', by Yuhai Tu, Thomas S. Shimizu, and Howard C. Berg, *Proc. Natl. Acad. Sci. USA* 105: 14855–60 (2008).

(6) 'Relationship between cellular response and behavioral variability in bacterial chemotaxis', by Thierry Emonet and Philippe Cluzel, *Proc. Natl. Acad. Sci. USA* 105: 3304–9 (2008).

# 10 Optimization of pathway

The chemotaxis signaling pathway has acquired a number of advantageous features, such as receptor clustering to amplify weak stimuli or precise adaptation to allow chemotaxis over a wide range of background concentrations. However, this is not the full story, as we can easily imagine that these are only the broad characteristics, but that there is a bit of fine-tuning in the details to further improve the performance. In this chapter we will focus on this fine tuning.

## 10.1 Optimal receptor-complex size

Previously, we discussed in detail the functional role of receptor clustering, and how this is achieved by receptor–receptor coupling. Despite all this insight, we can further ask what determines complex size: that is, the number of strongly cooperative receptors in a complex. Does it matter for the cell if five or ten receptors cooperate?

A detailed analysis of the FRET data from Fig. 2.7A shows that complex size increases with receptor methylation level (actually with the number of glutamine (Q) substitutions of glutamates (E) in genetically engineered cells). Since the methylation level reflects the ligand concentration the receptors are adapted to, this effectively means that the complex size increases with ligand concentration. This is illustrated in Fig. 10.1, showing inferred complex size with confidence intervals.

Variable complex sizes seem to make sense when considering what effects ligand input noise has on signaling: at low concentrations the relative ligand noise is the largest as the fluctuations in ligand concentration at the receptors increase with the background concentration (see Chapters 8 and 9 for a discussion of this Poisson noise). Hence, the amplification of weak stimuli should be reduced at small ligand concentrations as also noise is amplified. Such noise can ultimately saturate receptor signaling activity, rendering sensing useless.

However, there might be other determinants of complex size. Complex size could be limited by an imperfect physical clustering mechanism as proteins and lipids are soft materials, undergoing substantial thermal motion. Furthermore, larger complexes may not form due to the presence of other proteins in the membrane, which may constitute impurities in the receptor cluster. This view is supported by the observation that receptor-cluster associated proteins are relatively dynamic. Or, the increase in complex size with methylation may just be accidental: methylation removes charged glutamates on receptors. This may reduce

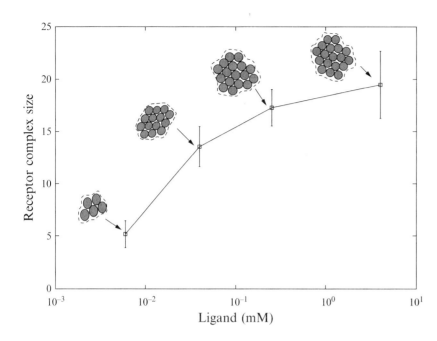

**Fig. 10.1** Analysis of dose-response curves from FRET in Tar-only cells indicates that complex size increases with receptor-modification level and hence effectively with attractant concentration. Data from Ref. (1).

electrostatic receptor-receptor repulsion, although the Deby–Hückel screening length inside the cell is only about 1 nm, which should efficiently neutralize any charges.

Here we explore the view that complex size is determined by functionality, and hence is optimal for sensing from an engineering perspective. We expect that a proper analysis would require the calculation of the *signal-to-noise ratio* (SNR), for which not only the effects of external ligand noise but also of internal signaling noise are required.

Fig. 10.2 summarizes some of the advantages and disadvantages of receptor clustering. On the one hand, more receptor cooperativity, i.e. larger complexes, amplify signals better in the absence of ligand input noise (top panels). A small stimulus such as a change in ligand concentration $\Delta c$ leads to a large activity response $\Delta A = (\partial A/\partial c)\Delta c$ (valid in the linear regime), where the derivative

$$\frac{\partial A}{\partial c} = N\bar{A}(\bar{A} - 1)\left(\frac{1}{K_D^{\text{off}} + c} - \frac{1}{K_D^{\text{on}} + c}\right) \qquad (10.1)$$

is proportional to the complex size $N$. (As usual this follow from MWC model for $N$ Tars with the derivative evaluated at adapted activity $\bar{A}$). On the other hand, random fluctuations in ligand concentration $\delta c$ also become amplified by the complex via $\delta A = (\partial A/\partial c)\delta c$ (bottom panels). Furthermore, the closer the proximity between receptors in a cluster, the larger the fluctuations in ligand concentration for the cell, because nearby receptor complexes measure previously bound ligand molecules due to rebinding (discussed in Chapter 8). Hence, clustering may render complexes highly prone to noise and reduce the cells signal processing capabilities. Indeed, sources of noise are ubiquitous in biological sensing.

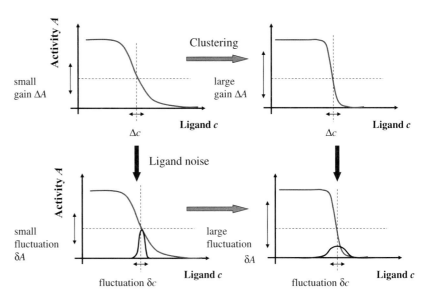

**Fig. 10.2** Schematic of dose-response curves of receptor activity. (top panels) Receptor cooperativity leads to steeper curves and hence larger amplification $\Delta A$ of a small stimulus (change in ligand concentration) $\Delta c$. (bottom panels) Noise in the stimulus, $\delta c$, represented by a peaked distribution, is amplified by receptor cooperativity as well, indicated by a fluctuation in activity $\delta A$. Receptor clustering, a potential side-effect of complex formation, leads to a further increase in input noise, shown by a broader distribution on the right. Redrawn from Ref. (2).

For simplicity, we focus here on the cell-external ligand and cell-internal methylation noise. Since the concentrations of methylation enzyme CheR and demethylation enzyme CheB are low, i.e. about a hundred copies per cell, the fluctuations in receptor methylation level are expected to be significant (see Fig. 9.8). In addition, there are internal random biochemical events further downstream of receptor signaling which we neglect.

Before embarking on an analysis of the SNR, it is instructive to know how the different noises in the pathway, external and internal, are processed by receptor complexes. We know that external ligand noise is amplified and signaling downstream of the complexes is not. However, the situation is less clear regarding internal receptor-complex noise, such as originating from receptor methylation.

The noise from random receptor-methylation and demethylation events was estimated as in Chapter 9 using the Langenvin approximation. Briefly, without loss of generality we use the simple linearized dynamics of the methylation level of a complex (see Eq. (9.16))

$$\frac{d(\delta M)}{dt} = \lambda_M \delta M + \eta_M,\tag{10.2}$$

where $\lambda_M = (N_R g_R + N_B g_B)\partial A/\partial M$ is the relaxation rate and $\eta_M$ is Gaussian white noise. (Note specifically that $N_R$ and $N_B$ is the number of tethered CheR and CheB enzymes per complex.) After Fourier-transforming we obtain for the noise power spectrum of the methylation level

$$\langle |\delta \hat{M}|^2 \rangle = \frac{Q_M}{\lambda_M^2 + \omega^2}\tag{10.3}$$

with noise intensity $Q_M = 2N_R g_R(1 - \bar{A})$ since each enzyme CheR and CheB contributes to the noise independently. To calculate the noise in methylation level $M$, we calculate the total variance of the methylation level, obtained by integrating over $\omega$

$$\langle(\delta M)^2\rangle = \frac{1}{\beta} \qquad (10.4)$$

with $\beta = \partial(\Delta F)/\partial M$ *independent of complex size* and $\Delta F$ the energy difference in Eq. (9.1). The conceptual reason for this $N$-independence is that a large complex has more enzymes bound to the receptors and hence suffers from larger noise than a small complex. However, a large complex also has an increased relaxation rate $\lambda_M$, restoring the average methylation level more quickly. Finally, due to noise propagation, Eq. (10.4) translates into the variance of the receptor complex activity. For simplicity, we consider here the instantaneous, unfiltered noise $\langle(\delta A)^2\rangle_M = (\partial A/\partial M)^2\langle(\delta M)^2\rangle$.

Next we estimate the noise contributions in the activity from the external ligand noise. Following the ideas from Chapters 8 and 9, the fluctuations in ligand concentration in a small volume near a receptor complex is related to the number of molecules $L$ inside this volume ($\sim s^3\bar{c}$ with $s$ the dimension of the volume), given by $\langle(\delta c)^2\rangle/\bar{c}^2 = \langle(\delta L)^2\rangle/\bar{L}^2 = 1/(s^3\bar{c})$, assuming the molecule number is Poisson distributed. However, this is the instantaneous uncertainty. If the cell instead has time $\tau$ available to average the stimulus, it can effectively make $D\tau/s^2$ measurements, leading to the reduced noise Eq. (8.3). This means that the uncertainty is the smaller the faster diffusion and the more spatial and temporal averaging the receptor complex is able to do. This translates into the (instantaneous) activity noise via $\langle(\delta A)^2\rangle_c = (\partial A/\partial c)^2\langle(\delta c)^2\rangle$. Due to amplification, this noise is proportional to the square of the complex size, $N^2$.

Hence, we established that external ligand noise is amplified, but not internal methylation noise. To continue with our investigation of an optimal complex size, we consider the SNR for sensing ligand stimuli by the whole cell for a fixed total number of receptors. Specifically, we assume $N_T$ identical receptors per cell, which are divided into $N_C$ smaller receptor signaling complexes of $N$ Tar receptors each with $N_T = N_C \cdot N$. We now calculate the $N$-dependent SNR for the total activity $A_T$ of the cell in response to a non-saturating change in ligand concentration $\Delta c$

$$\text{SNR} = \frac{\text{Signal}}{\text{Noise}} = \frac{\langle\Delta A_T\rangle_N^2}{\langle\langle\delta A_T^2\rangle\rangle_N}, \qquad (10.5)$$

where the *Signal* is defined by the squared-mean response of all receptors $\langle\Delta A_T\rangle_N^2 = \langle\sum_i^{N_T}\Delta A_i\rangle^2 = N_T\langle\Delta A\rangle^2$ to the stimulus, and the *Noise* is expressed by the mean-square deviation of the independently fluctuating receptor complexes $\langle\langle\delta A_T^2\rangle\rangle_N = \sum_j^{N_C}\langle\delta A_j^2\rangle = N_T N\langle(\delta A)^2\rangle$. As indicated by these formulas and illustrated in Fig. 10.3, the Signal and Noise behave very differently. The Signal is a coherent response of all complexes in a cell to a stimulus, while the Noise is due to incoherent fluctuations of signaling from the individual complexes. Note the Noise can have contributions from various sources. For methylation and ligand

A  Signal

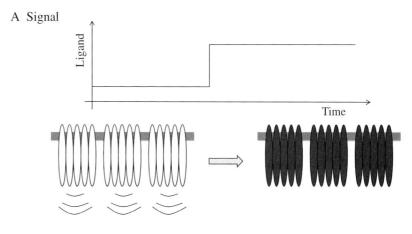

B  Noise

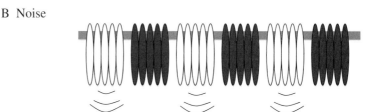

**Fig. 10.3** Schematics of the *Signal* (A) and the *Noise* (B). The Signal is a coherent response of all receptor complexes to a stimulus (change in ligand concentration). The Noise is due to incoherent signaling from complexes fluctuating between active and inactive conformations in absence of a stimulus.

noise we obtain $\langle (\delta A)^2 \rangle = \langle (\delta A)^2 \rangle_M + \langle (\delta A)^2 \rangle_c$. Plugged into Eq. (10.5), and using their respective $N$-dependencies, we obtain

$$\text{SNR} = \frac{N_T < \Delta A >^2}{N(\ll \delta A^2 \gg_M + \ll \delta A^2 \gg_c)} \sim \frac{N_T \cdot N}{const + N^2}. \qquad (10.6)$$

Fig. 10.4 shows a plot of this equation for cases with only internal noise, only external noise, and with both. Only when both external and internal noise are present, an optimal complex size arises which maximizes the SNR: in Eq. (10.6), starting from small complexes the SNR is proportional to the complex size, since the external noise is negligible to the internal noise in the denominator. Only when the external noise becomes comparable to the internal noise for larger complex sizes and hence amplification, the SNR reaches a maximum. For even larger complex sizes, the SNR declines again. The latter is the case since Signal and Noise have intrinsically different $N$-dependencies due to their coherent versus incoherent signaling behavior.

Importantly, if there was only external noise, the amplification of the noise is always stronger than the amplification of the stimulus due to their different $N$-dependencies. Hence, in this case complexes are not beneficial for the cell at all. In contrast, if there was only internal noise, there would be no penalty for larger and larger complexes (if the cell could make them and had enough available membrane surface area). Hence, in this case, the larger the complexes the better for the cell (disregarding slowing down of the response).

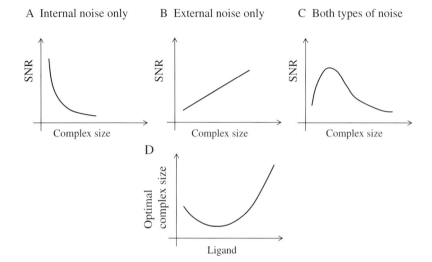

**Fig. 10.4** Signal-to-noise ratio (SNR) as a function of complex size for internal noise only (A), external noise only (B), and both types of noises (C). The optimal complex size, defined as the complex size, which maximizes the SNR ratio in presence of both noises, is plotted as a function of ligand concentration (D). Based on Ref. (2).

## 10.2 Optimal adaptation dynamics

The second example for discussing further optimization of pathway features concerns precise adaptation. Tethered-cell experiments show that the motor bias as a function of CheY-P concentration is remarkably sharp (switch-like) with a Hill coefficient above 10 (see Fig. 10.5A). This shows that adaptation does not only need to be precise but that the steady-state concentration of CheY-P needs to fall into the very narrow range of the sensitivity of the motor despite potentially large variations in chemotaxis protein concentrations. Hence, adaptation also needs to be precise in an absolute sense. (Later in this chapter we will soften this assumption a little.)

To achieve robust precision, a pathway should ideally be robust to concerted up- or down-regulation of some or all the proteins involved. Let $\{x_1(t), x_2(t), \ldots, x_N(t)\}$ be the time-dependent concentrations of the activated $N$ proteins of interest,[1] determined by

[1]See Ref. (4) for details.

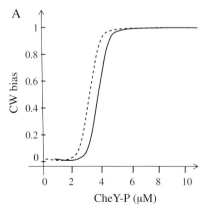

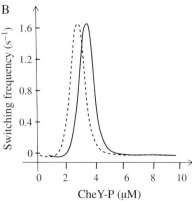

**Fig. 10.5** Motor behavior as a function of CheY-P concentration from tethered-cell assays. Shown are (A) the motor clockwise (CW) bias and (B) corresponding CW-CCW switching frequency for two slightly shifted sets of curves (solid and dashed lines). Based on data from Refs. (3) and (5).

$$\frac{dx_i}{dt} = F_i(\mathbf{x}(t)), \tag{10.7}$$

with the total protein concentrations $x^T = \{x_1^T, x_2^T, ..., x_N^T\}$ (active plus inactive) assumed constant on the time scale of adaptation. The stationary solution $\bar{\mathbf{x}}$ is obtained from the condition $F_i(\bar{\mathbf{x}}) = 0$. For this solution to be invariant against varying total protein concentrations, we require homogeneity of $F$ with respect to $x_i^T$,

$$F_i(\xi \mathbf{x}^T) = \xi^{\mu_i} F_i(x^T) \tag{10.8}$$

with $\mu_i > 0$. For instance, a $\xi$-fold change in transcriptional activity of all the proteins would result in $x^T = (x_1^T, ..., x_M^T) \to (\xi x_1^T, ..., \xi x_M^T)$.

The kinetics of the biochemical reactions of the chemotaxis pathway can be chosen in a way that homogeneity is fulfilled. However, since this makes a number of modeling assumptions, we step back and rather point out some examples contributing to this type of robustness. For instance, take the key response regulator, CheY. Since the concentration of CheY-P ($Y_p$) needs to fall within the steep sensitive region of the motor response for chemotaxis to work, we assume that its time evolution is given by the simple relationship

$$\frac{dY_p}{dt} = k_y \cdot T \cdot A \cdot (Y_T - Y_p) - k_{\bar{Y}} \cdot Y_p. \tag{10.9}$$

The first term on the right-hand side corresponds to phosphorylation of CheY (here modeled as $Y_T - Y_p$) with $Y_T$ the total CheY concentration. This phosphorylation is done by the active fraction $A$ of receptors with concentration $T$. The second term on the right-hand side corresponds to dephosphorylation of CheY-P. The steady-state concentration of CheY-P is hence given by

$$\bar{Y}_p = \frac{k_Y T A Y_T}{k_y T A + k_{\bar{Y}}}. \tag{10.10}$$

Upon a change of all protein concentrations by factor $\xi$, we can rewrite and obtain instead

$$\bar{Y}_p = \frac{k_Y(\xi T)A(\xi Y_T)}{k_y(\xi T)A + k_{\bar{Y}}} \sim \xi^2, \tag{10.11}$$

which forces $\bar{Y}_p$ to change up to a factor $\xi^2$ for a large dephosphorylation rate $k_{\bar{Y}}$. However, since the *E. coli* chemotaxis pathway has a phosphatase CheZ ($Z$) for CheY-P, we obtain instead

$$\bar{Y}_p = \frac{k_Y(\xi T)A(\xi Y_T)}{k_y(\xi T)A + k_{\bar{Y}}(\xi Z)} \sim \xi \tag{10.12}$$

with a reduced $\xi$ dependence, as one factor $\xi$ can be pulled out of the numerator and the denominator and thus cancel each other. If the upstream amount of active receptors, $AT$, were regulated to be independent of $\xi$, then $\bar{Y}_p$ would be completely independent of $\xi$. From this we derive the general principle that kinases may need a phosphatase to become robust. Interestingly, not all chemotaxis pathways in different bacterial species have dedicated phosphatases, and hence might require other pathway features to become robust.

A pathway should also be robust to perturbations in the amounts of individual protein components. To illustrate this type of robustness, we consider the simple model for precise adaptation based on Eq. (6.9), where the rate of change of the average receptor methylation level is given by (cf. Fig. 10.6A)

$$\frac{dM}{dt} = \tilde{N}_R \cdot R \cdot g_R \cdot (1 - A) - \tilde{N}_B \cdot B \cdot g_B \cdot A, \qquad (10.13)$$

now explicitly showing the dependence of the rate of methylation and demethylation on enzyme concentrations CheR ($R$) and CheB ($B$), e.g. with $N_R = \tilde{N}_R R$ the number of tethered CheR enzymes and similarly for the number of tethered CheB enzymes. (Note that in this simple model CheB does not need to be phosphorylated.)

Both Eq. (10.9) for the dynamics of CheY-P and Eq. (10.13) for the dynamics of the methylation level $M$ are coupled through the active receptor fraction $A$. Now, we consider these two equations at steady state and introduce small perturbations in $R \rightarrow \bar{R} + \Delta R$, $A \rightarrow \bar{A} + \Delta A$, and $Y_p \rightarrow \bar{Y}_p + \Delta Y_p$, keeping only linear terms in the perturbation. Removing $\Delta A$ in these equations, we can derive a relation how a perturbation $\Delta R$ leads to a perturbation $\Delta Y_p$,

$$\Delta Y_p = \frac{C}{\tilde{N}_R g_R \bar{R} + \tilde{N}_B g_B \bar{B}} \Delta R \qquad (10.14)$$

with $C$ a constant. If we however include that CheB requires phosphorylation to become catalytically active for demethylation, we obtain (cf. Fig. 10.6B)

$$\frac{dM}{dt} = \tilde{N}_R R g_R (1 - A) - \tilde{N}_B g_B (B_p)^{1+n} A, \qquad (10.15)$$

with $n > 0$ for $B_p$ cooperativity. Now we also require an equation for the rate of change of $B_p$

$$\frac{dB_p}{dt} = k_{Bp} A (B_T - B_p) - k_{\bar{B}p} B_p. \qquad (10.16)$$

where $B_T$ is the total CheB concentration and the two terms on the right-hand side correspond to activation by phosphorylation and deactivation by (auto)dephosphorylation. At steady state we solve this equation for $\bar{B}_p(A)$ and plug the result into Eq. (10.15) to eliminate the $B_p$-dependence. Again we linearize as before and now obtain for the perturbation in CheY-P

$$\Delta Y_p = \frac{C}{\tilde{N}_R g_R \bar{R} + \tilde{N}_B g_B [(\bar{B}_p)^n + \bar{A} \cdot \partial (\bar{B}_p)^n / \partial A]} \Delta R. \qquad (10.17)$$

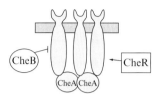

A   without CheB-P feedback

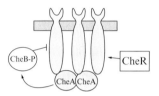

B   with CheB-P feedback

**Fig. 10.6** Schematic of adaptation module without (A) and with (B) negative feedback by CheB phosphorylation by active receptors. Note that (A) still contains basic negative feedback as demethylation occurs only for active receptors. Module (B) corresponds more closely to actual chemotaxis pathway. Pathways (A) and (B) were measured in Ref. (6).

Since $\partial(\bar{B}_p)^n/\partial A$ in the denominator is positive, the derivative reduces the perturbation $\Delta R$ induces on the CheY-P concentration. Hence, such an additional negative feedback loop can improve the robustness to perturbations and noise.

So far we have assumed that the cell requires absolute adaptation of the CheY-P concentration so that the steady-state concentration of CheY-P falls within the narrow range of the steep motor bias curve. This constraint may, however, not be that strict. For a prolonged change in CheY-P level, the motor may be able to shift its response threshold illustrated by solid and dashed lines in Fig. 10.5. This can potentially be achieved if the number of FliM proteins per motor (and hence their cooperativity) changes. For instance, the trunover rates of FliM proteins in the motor may depend on the CheY-P level.

There is another important aspect about the dynamics of adaptation. Specifically, the dynamics need to reflect the temporal changes of the chemicals in the environment and the swimming behavior of the cells. At a very basic level we would expect that the runs should be long (however, not longer than the time scale of rotational diffusion which eventually bends the run path away from a straight line) and that tumbles are short (just long enough to randomize the cell's swimming direction).

To investigate this, we compare adaptation model Eq. (10.13) without negative CheB-P feedback with adaptation model Eq. (10.15) with such feedback. As shown in the sample time course of Fig. 10.7, the dynamics of adaptation are quite different between the two models when considering a step increase and decrease of attractant. Without feedback the dynamics of adaptation to addition of attractant is similar (although of opposite sign) to the dynamics of adaptation to removal of attractant. Hence, the duration of tumbling is very long. However, with feedback (here $n = 1$), the dynamics become very asymmetric, characterized by long durations of runs and very short durations of tumbles. Importantly, the model with feedback is much closer to experimental FRET time-

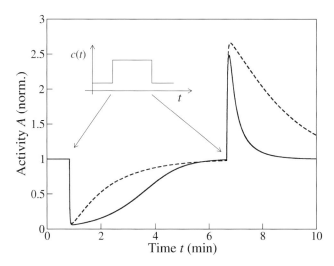

**Fig. 10.7** Comparison of adaptation models with (solid line) and without (dashed line) negative feedback by phosphorylation of CheB. Shown are time courses of adaptation to a step increase and subsequent decrease of attractant (inset). Redrawn from Ref. (6).

course data. From the perspective of a swimming bacterium, feedback allows for more efficient reorientation of the cell. This in return might facilitate the quick escape of the cell from toxic chemicals.

Generally important features for the cell, such as precise adaptation, need to be implemented very robustly (here by integral feedback). This robustness allows the feature to work independent of certain perturbations in the pathway. For instance, we learned in Chapter 6 that the precision of adaptation is robust to concerted fluctuations in CheR and CheB expression level. However, there will always be perturbations to which the feature is not robust, e.g. if only CheR expression goes up and not the expression level of CheB. For these circumstances we can imagine that some additional fine tuning is required, which was illustrated in this chapter. However, evidence for such subtleties is difficult to come by, often requiring detailed biochemical studies in live cells, as *in vitro* studies on cell extracts may show artifacts.

# Chapter summary

- To increase the fitness of the cells, their sensing and signaling are expected to be optimized, i.e. the work as good as possible given energetic and physical constraints.

- Analysis of FRET dose-response curves of chemotaxis signaling shows that complex size increases with receptor methylation level. Due to adaptation, the methylation level corresponds to increasing ligand concentration, so that complex size appears to increase of background stimulus. This makes sense, as at larger ligand concentration the relative ligand noise decreases, making amplification of noise less of an issue, but a quantitative analysis is required.

- To analyse this we calculated the signal-to-noise ratio (SNR) for the signaling activity in a cell $A_T$ with a fixed number of receptors $N_T$ but variable complex size $N$. We found that the SNR is given by

$$\text{SNR} = \frac{\text{Signal}}{\text{Noise}} = \frac{\langle \Delta A_T \rangle_N^2}{\langle \langle \delta A_T^2 \rangle \rangle_N} \sim \frac{N_T \cdot N}{const + N^2}, \tag{10.18}$$

where $\Delta A_T$ is the coherent response of the activity from all complexes to a ligand step change and $\delta A_T$ is the incoherent noise from unstimulated independent receptor complexes. The right-hand side of the above formula has a maximum, indicating an optimal complex size $N^*$.

- Pathways should ideally be robust towards fluctuations in protein number. Given general pathway dynamics

$$\frac{dx_i}{dt} = F_i(\mathbf{x}(t)), \tag{10.19}$$

with the stationary solution $\bar{\mathbf{x}}$ obtained from the condition $F_i(\bar{\mathbf{x}}) = 0$. For this solution to be invariant against varying total protein concentrations, we require homogeneity of $F$ with respect to $x_i^T$,

$$F_i(\xi\mathbf{x}^T) = \xi^{\mu_i} F_i(x^T). \tag{10.20}$$

The chemotaxis pathway can be designed to have these features, e.g. by having a phosphatase CheZ. In addition to global changes in expression level of all pathway components, the pathway should also be robust to individual protein fluctuations, e.g. that the CheY-P level is relatively robust towards changes in enzyme CheR for specific realizations of the adaptation dynamics and feedback mechanisms.

- In the adaptation dynamics, the rate of demethylation may look very different from the rate of methylation. While CheR simply tethers to the receptor to increase its local concentration, CheB may act cooperatively. Such enzyme cooperativity may help with the aforementioned robustness, as well as leading to long runs while tumbles remain very short. This guarantees that tumbles are just long enough to randomly select a new direction while not wasting valuable time for the swimming bacterium.

# Further reading

(1) 'Variable sizes of *Escherichia coli* chemoreceptor signaling teams', by Robert G. Endres, Olga Oleksiuk, Clinton H. Hansen, Yigal. Meir, Victor Sourjik, Ned S. Wingreen, *Mol. Syst. Biol.* 4: 211 (2008).

(2) 'Optimal receptor-complex size determined by intrinsic and extrinsic noise' by Gerardo Aquino, Diana Clausznitzer, Sylvain Tollis, and Robert G. Endres, *Phys. Rev.* E 83: 021914 (2011).

(3) 'An ultrasensitive bacterial motor revealed by monitoring signaling proteins in single cells', by Philippe Cluzel, Michael Surette, and Stanislav Leibler, *Science* 287: 1652–5 (2000).

(4) 'Design principls of a bacterial signalling network', by Markus Kollmann, Linda Lovdok, Kilian Bartholome, Jens Timmer, and Victor Sourjik, *Nature* 438: 504–7 (2005).

(5) 'Adaptation at the output of the chemotaxis signalling pathway', by Junhua Yuan, Richard W. Branch, Basarab G. Hosu, and Howard C. Berg, *Nature* 484: 233–6 (2012).

(6) 'Chemotactic response and adaptation dynamics in Escherichia coli', by Diana Clausznitzer, Olga Oleksiuk, Linda Lovdok, Victor Sourjik, and Robert Endres, *PLoS Comput. Biol.* 6: e1000784 (2010).

# "Seeing like a bacterium"

**11**

Most experiments are done under idealized laboratory conditions. Although we have learned a lot from these about *E. coli* chemotaxis, we know surprisingly little about the natural environment of this bacterium. We know what chemicals bind to the receptors, but what are the chemical gradients bacteria typically encounter? What if we could "see" like a bacterium to learn more? To start addressing these questions, we should know a bit more about the life-style of this bacterium.

## 11.1 Typical chemical gradients

*E. coli* lives in the large intestine, along with hundreds of other bacterial species at high cell densities near the mucosal epithelial surface. Note the epithelium is a skin-like sheet of cell layers lining our intestines and provides an important barrier between the interior and exterior of our body. At this rather inaccessible place, *E. coli* synergistically helps us digest food and obtain vitamins.

We expect this chemical environment to be highly dynamic with complex spatio-temporal gradients from partially digested food and host secretions. Although the exact role of chemotaxis for non-pathogenic strains of *E. coli* is unknown, the chemotactic ability may provide a selective advantage to enable cells to remain in their niche despite the highly dynamic turnover of mucous in the gut and the severe competition from other bacterial species (see Fig. 11.1). Alternatively, chemotaxis may have advantages for the survival of the bacteria when living outside the host, where nutrients are expected to be much more sparse.

What type of chemical gradients might bacteria encounter? A number of characteristic chemical concentration profiles are shown in Fig. 11.2 and originate as follows. If a fixed number of ligand molecules is deposited in solution, free diffusion leads to a Gaussian-shaped gradient (see Chapter 3). If diffusion occurs from a point source, which is emitting ligand molecules at a constant rate, the result is a hyperbolically shaped gradient. Another type of gradient, a linear gradient, occurs between points of constant emission and absorption. If instead ligand molecules are emitted from a constant source in presence of homogeneous degradation throughout the medium, an exponential gradient is the result. One might argue that local high concentrations of chemicals with bacteria around almost everywhere might lead to exponential gradients, although one cannot be really sure and it is uncertain whether well-defined gradients exist at all.

$c(x,t)$

**Fig. 11.1** Illustration of the dynamic chemical environment of bacteria. Shown are different co-existing bacterial species, often competing for the same niche, leading to an inhomogeneous concentration profile.

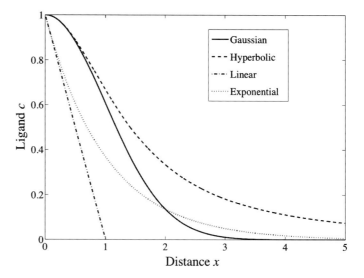

**Fig. 11.2** Chemical concentration profiles with different properties. The Gaussian profile decays faster than the exponential profile. The exponential profile decays faster than the hyperbolic profile, which has a power-law behavior. The linear gradient unrealistically decays to zero, but may be a good approximation over short distances.

To learn more about the bacteria's typical chemical environments, we go a different route and make the assumption that the chemotaxis sensory pathway is optimized by evolution to the gradients bacteria typically encounter and conquer. So when we know a lot about a sensory system, such as bacterial chemotaxis, we can try to predict the chemical environments by reverse engineering.

## 11.2   Weber's law

The first approach we take is to consider the phenomenological Weber's law, which most sensory systems fulfil to a good approximation, including our own senses. This "law" says that the smallest detectable stimulus, e.g. change in ligand concentration $\Delta c$, increases linearly with the background concentration $\bar{c}$; that is,

$$\Delta c = K \cdot \bar{c}, \tag{11.1}$$

where $K$ is the constant Weber fraction. In other words, the sensory system encodes contrast $\Delta c/c_0$. When applied to our visual system, this would allow us to exploit the statistical similarity of the light intensities in natural scenes. The light intensity reflected from objects relative to the background is largely invariant. Hence, when using Weber's law the sensing and encoding of about ten orders of naturally occurring light intensities is made easier. We humans experience Weber's law every day. In a dark environment we are very sensitive detecting nearly single photons arriving on our retina but not in bright day light. In a loud environment we need to shout so someone nearby can hear us, but if in a very quiet environment it is sufficient to wisper.

In its integrated form, Weber's law predicts an internal representation of the stimulus, perception $P_e$, to increase logarithmically with the stimulus $c$

$$P_e \sim \ln c. \qquad (11.2)$$

This relationship, known as the Weber–Fechner law, leads to a logarithmic stimulus compression and hence a large dynamic range. In general, the detailed understanding of the molecular origin of these laws is nearly impossible. For instance, the photoreceptors of the vertebrate eye involve up to nine different mechanisms for adaptation, not including downstream neural processes. Fortunately, the bacterial chemotaxis system is much simpler and fully characterized, so we can try to understand the molecular emergence of these laws, their limits, and what they tell us about the chemical gradients.

To derive Weber's law we examine how a barely detectable stimulus is transmitted by the receptors. For this purpose we expand the receptor-complex activity for a small change in ligand concentration via (see Chapter 9)

$$\Delta A = \left. \frac{\partial A}{\partial (\Delta F)} \right|_{\Delta \bar{F}} \cdot \left. \frac{\partial (\Delta F)}{\partial \ln c} \right|_{\bar{c}} = \alpha \cdot k \cdot \frac{\Delta c}{\bar{c}}, \qquad (11.3)$$

where $\Delta \bar{F}$ is the adapted free-energy difference of a complex corresponding to the adapted activity $\bar{A}$, $\alpha = \bar{A}(\bar{A} - 1)$, and $k$ a prefactor to be discussed below. Let $\Delta A$ be the smallest activity change (threshold response), which leads to a noticeable response of the motor. This defines $\Delta c$ as the smallest detectable stimulus (threshold stimulus).

Importantly, in Eq. (11.3) the prefactor $k = N[\bar{c}/(\bar{c} + K_D^{\text{off}}) - \bar{c}/(\bar{c} + K_D^{\text{on}})]k_B T$ is proportional to $N$ and approximately constant for $K_D^{\text{off}} < < \bar{c} \ll K_D^{\text{on}}$. (For simplicity, we neglected ligand binding to other types of receptors here.) As a result, we successfully derived Weber's law (Eq. (11.1)) with Weber fraction $K = \Delta A/(\alpha k)$. For ligand concentrations significantly below $K_D^{\text{off}}$ and above $K_D^{\text{on}}$, $K$ is not constant anymore and Weber's law breaks down. Specifically, for $c \ll K_D^{\text{off}}$ the receptors do not bind ligand molecules, and we expect that the smallest detectable change in ligand concentration approaches a constant: that is, for very small ligand concentrations the cell requires more addition of ligand to launch a significant response than predicted by Weber's law. In contrast, for $c \gg K_D^{\text{on}}$ Weber's law may break down because of two reasons: (1) receptors are precisely adapted to the background concentration (sufficient number of methylation sites available on receptor) but no longer respond to further increases in concentration; (2) adaptation to the background concentration is imprecise, as receptors run out of methylation levels.

The range of validity of Weber's law can be tested by FRET experiments (Chapter 2). For a given background concentration, the initial responses to increasing stimuli (step changes of attractant) can be measured and plotted as dose-response curves (Fig. 11.3A). Now the threshold stimuli $\Delta c_t$ need to be inferred. This can be done by considering a threshold response $\Delta A$ (assumed the same for all curves and hence represented by horizontal lines for addition and removal of attractant, respectively). The values for $\Delta c_t$ can be inferred from the corresponding

**Fig. 11.3** Validity of Weber's law in bacterial chemotaxis signaling. (A) Dose-response curves for four different background concentrations. Lower branches correspond to addition of attractant, upper branches to removal of attractant. The smallest detectable stimulus (threshold stimulus $\Delta c_t$) is indicated by exemplar arrows and is defined as the change in ligand concentration, which leads to a noticeable change in activity (horizontal lines). Redrawn from Ref. (7). (B) Threshold stimulus as a function of background concentration. The linear behavior breaks down at low and high ligand concentrations. Ligand dissociation constants of Tar receptor are so shown.

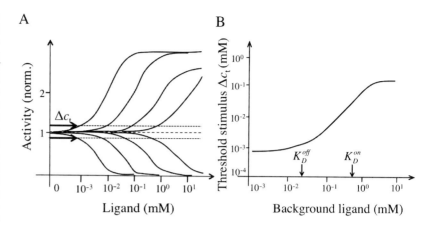

ligand concentrations (for an example, see the arrows in Fig. 11.3A). The exact value of $\Delta A$ should not matter as long as we are in the linear response regime of the receptor activity. From this we can plot $\Delta c$ over $\bar{c}$, and thus determine the range of background concentrations over which the plot is linear, and hence follows Weber's law. The result of such an analysis is shown in Fig. 11.3B, demonstrating that the threshold stimulus is indeed approximately linear for background concentrations between the two ligand dissociation constants of the Tar receptor. Note that the linear range is extended beyond $K_D^{on}$ by an order of magnitude due to a concomitant increase in receptor complex size (Chapter 10). There is also another linear regime with a different slope at higher concentrations, as Tsr receptors can bind this ligand as well, albeit with lower affinity.

So far the threshold response $\Delta A$ was rather arbitrary and mainly determined by the experimental noise in the FRET data, limiting us to resolve small stimuli in the noisy activity measurement. However, we can also go a different route and estimate $\Delta A$ directly. Put simply this $\Delta A$ needs to be larger than typical fluctuations in the activity $\delta A$ from noise in the sensory pathway. Using the approach from Chapters 9 and 10, $\Delta A$ can thus be calculated from external and internal noise sources.

## 11.3    Perception

The Weber–Fechner law is the integrated version of Weber's law and suggests that there is an internal representation of the stimulus called perception $P_e \sim \ln(c)$. What is this internal representation for chemotaxis? We cannot simply integrate Eq. (11.3) since derivative $\partial A/\partial(\Delta F)$ (or $\alpha$) depends implicitly on the ligand concentration (via $\bar{A}$). However, we can integrate $\delta(\Delta F) = k \cdot \delta c/c$, which shows that the perception is given by the receptor-complex energy difference

$$\Delta F \approx \Delta E(M) + N k_B T \ln \left( \frac{c}{K_D^{\text{off}}} \right). \tag{11.4}$$

Comparison with the exact expression of the energy difference, Eq. (5.27), shows that this, once more, is valid for $K_D^{\text{off}} \ll c \ll K_D^{\text{on}}$. Noticeable in the plot in Fig. 11.4A is a perception curve for each methylation level. When receptors are adapted to activity $\bar{A}$, the corresponding energy difference is $\Delta \bar{F}$ for a specific methylation level. Sudden changes in the ligand concentration move the perception along a methylation-level specific curve, whereas adaptation shifts the perception vertically to a new curve for a different methylation level.

What does the perception mean in terms of typical chemical gradients? For this purpose we consider a moving bacterium in different hypothetical gradients (see Fig. 11.4B) and compare how the cells would perceive such encountered stimuli (inset). The result is that when a cell moves

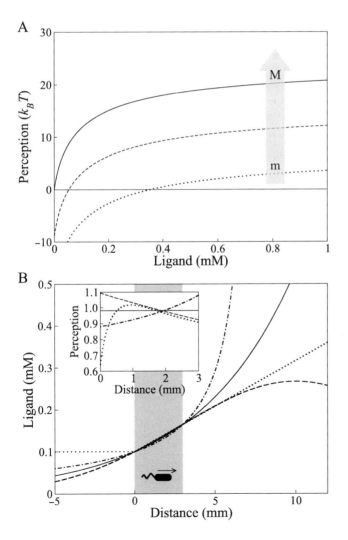

Fig. 11.4 Perception in chemotaxis. (A) Perception as predicted by Weber's law is represented in bacterial chemotaxis by the energy difference between the *on* and *off* states of receptor complexes. There is a different perception curve for each receptor methylation level. (B) Different spatial concentration profiles, coinciding in the gray region, leads to different perceptions for a bacterial cell. A cell is swimming with constant velocity to the right, up the gradients (solid lines). Exponential gradients (solid line) are special, as they lead to a constant perception (inset).

with constant velocity, only the exponential spatial gradient is perceived as constant in time. Such a constant perception seems advantageous to the cell; adaptation neither is able to eliminate the response, nor saturates the receptor activity. This can be seen as follows: the moving cells effectively experience an exponentially increasing gradient in time, e.g. $c(t) = c_0 e^{rt}$. For suitable attractant concentrations, the energy difference is given by Eq. (11.4), which now changes in time like

$$\frac{d(\Delta F)}{dt} = \frac{\partial(\Delta F)}{\partial M}\frac{\partial M}{\partial t} + \frac{\partial(\Delta F)}{\partial \ln c}\frac{d \ln c}{dt}. \tag{11.5}$$

Using our model for precise adaptation (Chapter 6), we obtain the following rate of change of the free-energy difference

$$\frac{d(\Delta F)}{dt} = -\frac{\Delta F}{\tau} + br \tag{11.6}$$

with appropriately defined constants $\tau$ and $b$. Eq. (11.6) is solved by

$$\Delta F(t) = \tau br \left(1 - e^{t/\tau}\right) \to \, \sim r \tag{11.7}$$

with the right-hand side approached after an initial transient for $t \gg \tau$. Hence, we obtain a time-independent change in energy difference and hence a perception proportional to the ramp rate $r$. Other types of gradients lead to either increasing or decreasing perceptions with time.

## 11.4   Fold-change detection

Weber's law deals with smallest detectable stimuli, which are important early guiding cues for any biological system. Via integration the law directly led us to the concept of perception and exponential gradients as ideal environments for bacteria. However, we can image that bacteria are often exposed to strongly fluctuating ligand concentrations, which can lead to large responses in signaling activity. Hence, it might be beneficial for the bacterium to have the validity of Weber's law extended to arbitrary stimuli, called *fold-change detection*.

   In fold-change detection any cellular response depends only on relative changes in ligand concentration, $\Delta c/\bar{c}_0$. As a result, we can scale the concentration by a positive factor $c \to \gamma c$, implying both $\Delta c \to \gamma \Delta c$ and $\bar{c} \to \gamma \bar{c}$. This leaves the relative change in ligand concentration invariant. Why is this relevant to cells? Generally, chemical concentration profiles are linear in the source strength. As an example, consider the exponential and Gaussian gradient in 1D

$$c_{\text{exp}}(x) = c_0 e^{-\lambda x} \tag{11.8}$$

$$c_{\text{Gauss}}(x, t) = \frac{c_0}{\sqrt{4\pi D t}} e^{\frac{-(x-x_0)^2}{4Dt}} \tag{11.9}$$

with $c_0$ the source strength, $\lambda$ the decay length of the exponential gradient, and $x_0$ the average of the Gaussian profile. Fold-change detection would allow cells to swim up these chemical concentration profiles as well as scaled versions of the profiles, extending the working range of the sensory system to a large number of profiles irrespective of overall magnitude of concentration.

Technically, the question of fold-change detection has a two parts, illustrated in Fig. 11.5. First, the response to sudden changes in ligand concentration must only depend on relative concentration changes. (This is what we have discussed so far.) Second, also the adaptation times must be the same to achieve a full collapse of response time courses.

As we will see now, bacterial chemotaxis can do fold-change detection. Let us consider an arbitrary response of the receptor-complex energy difference of initially adapted receptors

$$\Delta(\Delta F) = \Delta F(\bar{c} + \Delta c) - \Delta F(\bar{c})$$
$$= N \ln \left( \frac{1 + \frac{\Delta c}{\bar{c}} \frac{1}{1 + K_D^{\text{off}}/\bar{c}}}{1 + \frac{\Delta c}{\bar{c}} \frac{1}{1 + K_D^{\text{on}}/\bar{c}}} \right)$$
$$\rightarrow N \ln \left( 1 + \frac{\Delta c}{\bar{c}} \right), \tag{11.10}$$

where the last line is approximately valid the usual concentration range $K_D^{\text{off}} \ll c \ll K_D^{\text{on}}$. (Note for simplicity we again assumed a single receptor type, e.g. $N$ Tar receptors.) Eq. (11.10) shows indeed that response only depends on $\Delta c/\bar{c}$ as predicted by fold-change detection. This behavior is illustrated in Fig. 11.6 using our (by now well-known) dose-response curves, thus extending Weber's law to arbitrary responses.

Fig. 11.6 is about response amplitudes. What about adaptation times - do they also solely depend on relative changes in ligand concentrations? Eq. (11.10) shows the applicability of fold-change detection to changes in receptor-complex energy and hence signaling activity. In precise adaptation the rates of receptor methylation and demethylation depend only on activity. Hence, we expect that fold-change detection extends to adaptation times as well, although adaptation times are generally quite variable (see the discussions in Chapters 6 and 9).

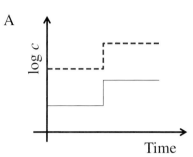

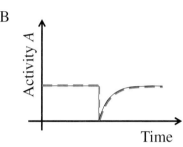

**Fig. 11.5** Concept behind fold-change detection. (A) A stimulus (black line) multiplied by a constant factor produces a shifted stimulus (gray line) on log scale. Note ligand concentration $c$ is made unitless here. (B) Both stimuli produce same activity response, indicating that only fold changes are detected. Based on Ref. (5).

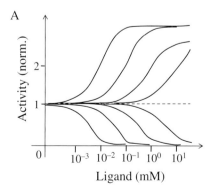

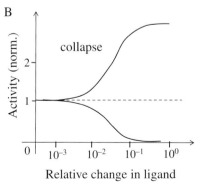

**Fig. 11.6** Fold-change detection in bacterial chemotaxis signaling. (A) Dose-response curves for four different background concentrations collapse onto single curve when plotted as a function of relative change in ligand concentration (B). Redrawn from Ref. (7). Note that the, zero background curve cannot be scaled this way.

## 11.5   Matching relations

Exponential gradients are special to bacteria, and may even represent typical gradients bacteria encounter—on average. But what about distributions of ligand concentration cells experience? This is a tricky question, since it is impossible to follow a cell around in a natural environment and record what it 'sees'. However, we might be able to exploit that sensory systems ought to be evolved to typical stimuli cells encounter. If not, the cell misses important information. The cell should also not be able to sense stimuli that it never encounters, as this wastes the cell's resources. Hence, we may expect to find a matching relation between typical stimuli and the corresponding cellular responses.

Since the resources of a cell are limited by energy constraints and noise, we expect that a cell tries to maximize the information transmission between input stimuli and activity outputs. Luckily, this transmission from inputs to outputs is very much accessible from FRET dose-response curves (Chapter 2). As these are recorded for specific receptor modification states, they correspond to the sensing of cells when adapted to certain background concentrations.

Fig. 11.7 illustrates how such a matching relation might work. Shown are a distribution of inputs, an input-output relation (dose-response curve), and a distribution of outputs. What are the relations to each other? Specifically, how should the receptor weigh the inputs that they are best represented by a finite range of cell-internal output states without wasting resources? Intuitively it is clear that inputs should be encoded such that each output level is used with equal probability. As a result

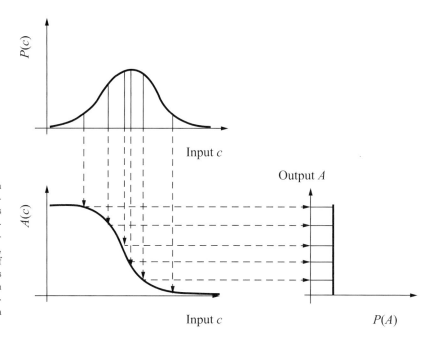

**Fig. 11.7** Maximizing information transmission predicts a matching relation between the distribution of inputs $P(c)$, the input-output relation (dose-response curve) $A(c)$, and the distribution of outputs $P(A)$. Without noise, equal areas under the distribution of inputs are mapped onto equal areas under the distribution of outputs with the input–output relation the cumulative distribution of inputs. Based on Ref. (2).

highly probable inputs are better resolved and represented by outputs than unlikely input values which hardly ever occur. This procedure predicts that the input–output relation is simply the cumulative distribution of the various inputs, because equal areas (probabilities) under the output distribution correspond to equal areas under the input distribution. This line of reasoning works indeed as has been demonstrated for multiple biological systems.

What happens if the input–output relation is corrupted by cell-internal noise as illustrated in Fig. 11.8? For such cases, we need a mathematically suitable measure for the information transmission. A measure describing the information contained in the output about the input (or vica versa) is the *mutual information* borrowed from communications research (Chapter 4). For one input variable ($c$) and one output variable ($A$), the mutual information is given by the functional

$$I[c, A] = \iint dc\, dA\, p(c, A) \log_2 \frac{p(c, A)}{p(c)p(A)}$$
$$= \int dc\, p(c) \int dA\, p(A|c) \log_2 \frac{p(A|c)}{p(A)}, \qquad (11.11)$$

where $p(c)$ is the probability distribution of input ligand concentrations and $p(A)$ is the probability distribution of output activities. Furthermore, $p(c, A)$ is the joint probability distribution of inputs and outputs, and $p(A|c)$ is the conditional probability distribution for output $A$ given input $c$. For the second line in Eq. (11.11) we rewrote the joint probability in terms of the conditional probability $p(c, A) = p(A|c)p(c)$. Eq. (11.11) describes a measure in bits (due to $\log_2$) and can be considered the logarithm of the numbers of states in the input that lead to distinguishable outputs, given the joint distribution and noise.

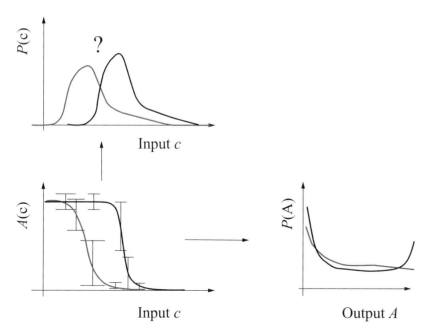

**Fig. 11.8** Measured dose-response curves (shown as schematics) with error bars for different modification states can be used to predict the distribution of inputs and outputs. Shown are two curves corresponding to two different receptor methylation levels. With cell-internal noise the distribution of outputs is bimodal.

The mutual information can generally not be calculated analytically. However, an exception is the special (but nevertheless useful) case that the internal noise is small and Gaussian distributed.[1] In this case, the input–output relation is given by the conditional probability distribution

$$p(A|c) = \frac{1}{\sqrt{2\pi\sigma_A^2(c)}} e^{-\frac{[A-\bar{A}(c)]^2}{2\sigma_A^2(c)}}, \qquad (11.12)$$

where $\bar{A}(c)$ is the average input–output relation corresponding to a dose-response curve, and $\sigma_A^2(c)$ is the variance of the output for a given input. The latter can be represented by error bars on the average dose-response curves (see Fig. 11.8). In this case we obtain for the mutual information the expression

$$I[c,A] = -\int d\bar{A}\, p(\bar{A}) \log_2 \sqrt{2\pi e \sigma_A^2}\, p(\bar{A}). \qquad (11.13)$$

Hence, to calculate the mutual information all we need are the distribution of average outputs $p(\bar{A})$ and the variance of outputs.

However, we do not necessarily want to calculate the mutual information. Instead we want the distribution of outputs, which will in return provide the desired distribution of inputs. Hence, we maximize the mutual information by calculating the functional derivative with respect to the output distribution and set the derivative to zero. Such an optimization procedure was outlined in Chapter 7 and yields

$$p^*(\bar{A}) = \frac{1}{Z} \frac{1}{\sigma_A(c(\bar{A}))}, \qquad (11.14)$$

where $Z$ is a normalization constant so that the integral over the probability distribution of the outputs is equal one (see also Appendix A). To obtain the desired distribution of inputs which maximizes the mutual information, we use conservation of probability

$$p^*(c) = p^*(\bar{A}(c)) \frac{d\bar{A}}{dc}, \qquad (11.15)$$

where $d\bar{A}/dc$ is the slope of the input-output relation. Eq. (11.14) indicates that output levels corrupted by a lot of noise (large $\sigma_A$) are used less frequently as they are unreliable. Since the activity is the noisiest in the steep region of the dose-response curve the resulting output distribution is bimodal (Fig. 11.8).

By combining Eqs. (11.14) and (11.15) we further obtain the distribution of inputs

$$p^*(c) \sim \frac{1}{\sigma_A(\bar{A}(c))} \frac{d\bar{A}}{dc}, \qquad (11.16)$$

which means that the input–output relation is the steeper the noisier the output levels to allow the cell to resolve noisy outputs. Based on these arguments we expect that if the input stimuli are noisy as well, the input–output relation should become shallower, thus allowing the cell to resolve different input signals. This is indeed the case, but the calculation is a little more complicated and hence is not shown here.

Let us briefly consider the special case when all output levels have the same noise. Now, using Eq. (11.15) we obtain $p^*(c) \sim d\bar{A}/dc$ so that the optimal input-output relation is given by the cumulative distribution of the inputs $\bar{A}(c) = \int_0^c dc' dc' \, p(c')$. This is equivalent to the case without noise. Furthermore, using Eq. (11.14) the distribution of outputs $p^*(\bar{A})$ is just uniform, so that all output levels are used with equal probability. This finding was already illustrated in Fig. 11.7, but now follows from mathematical derivation.

So far we have not mentioned the actual values of the mutual information, but have instead focused on the prediction of the distributions of inputs and outputs. However, given Eq. (11.14) and internal noise, we can actually calculate the maximal amount of transmitted information. The result is given by

$$I^* = \log_2(Z/\sqrt{2\pi e}), \tag{11.17}$$

which indicates that the cell can resolve $2^{I^*}$ different input states.

## 11.6   Predicting typical stimuli

We now return to bacterial chemotaxis with the aim of predicting the distribution of input ligand concentrations that bacteria see when adapted with a specific receptor methylation level. For this we can take FRET dose-response curves from Fig. 2.7A (see also illustrative examples in Fig. 11.8, lower-left panel). Although such measurements were conducted on many cells simultaneously, we can estimate the activity noise from repeated measurements of the curves. This activity noise can then be used to predict the distributions of inputs and outputs. Such an analysis returns the following. First, we find that the resulting outputs are bimodal for higher methylation states. We conclude that due to the increased steepness of these dose-response curves, intermediate activity levels appear too noisy for reliable sensing. Second, the distributions of inputs are log-normal to a very good approximation (normal on a log scale; see Fig. 11.9). If we scale them by their average ligand concentration, the distributions of inputs for different methylation levels collapse onto a single curve (see Fig. 11.9, inset). This validates Weber's law, since the width of the distributions is proportional to their average.

Taken together, the analysis of this chapter suggests that exponential gradients with log-normally distributed ligand concentration are special for bacteria. Since exponential gradients originate from constant chemical sources and homogeneous degradation, this may suggest highly localized nutrients with degradation by surrounding bacteria. We have thus obtained a glimpse into the secret lives of these bacteria and their hidden environments. Furthermore, the general applicability of information theory points towards a universal principle in biology across all scales. Biological systems seem to be about reliable sensing and information transmission.

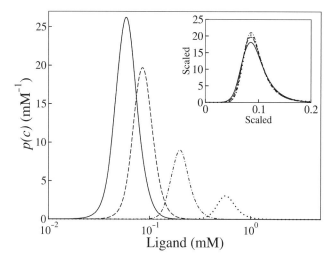

**Fig. 11.9** Log-normal distributions of the predicted optimal distributions of inputs for different background concentrations (main panel) and their collapse by scaling with the average values (inset).

## Chapter summary

- The bacterium *E. coli* can live in our intestines to help us digest food. Due to inaccessibility of this environment by experiments, little is known about the chemical environments that bacteria typically encounter. Ideally, we would like to follow bacteria around and to "see" like a bacterium. Although impossible, this can indirectly be achieved by re-engineering the chemical environments from all the information we have about the chemo-sensory system. The assumption is that the sensory system has been optimized for typical environments.

- The Weber's law says that the smallest detectable stimulus, here change in ligand concentration $\Delta c_t$, scales linearly with the background stimulus $\bar{c}$ via

$$\Delta c_t = K\,\bar{c}, \tag{11.18}$$

where $K$ is the constant Weber fraction. This constant can easily be calculated by expanding the activity round the adapted activity in terms of a small change in ligand concentration, indicating that Weber's law is valid between the *off* and *on*-state ligand dissociation constants of each receptor species with ligand binding capability. The Weber fraction depends on numbers of each receptor species in the complex, and hence may vary for different concentration ranges. Note, Weber's law appears to even hold for larger stimuli, and is then called fold-change detection.

- The integrated version of Weber's law is the Weber–Fechner law, which says that there is a quantity called perception

$$R \sim \ln(c), \tag{11.19}$$

which scales logarithmically with the stimulus (here concentration $c$). In bacterial chemotaxis, this perception is given by the receptor-complex energy. Due to the logarithmic dependence, predicted typical concentration gradients are exponential in space. For a swimming bacterium these lead to a constant perception. Such stimuli do not saturate the response.

- Assuming that cells maximize information flow in the sensory system and downstream pathway, information theory can be used to derive matching relations between the distribution of inputs (typical chemical stimuli), the input–output relation (dose-response curve), and the distribution of outputs (signaling activity). The derivation involves maximizing the mutual information between inputs and outputs with respect to the quantity of interest. If we know dose-response curves from experiment, we can derive both the optimal distributions of inputs and outputs.

- For small Gaussian-distributed internal noise, the predicted distribution of inputs can be calculated analytically. The result shows that the input–output relation is the cumulative distribution of the inputs, and that it is the steeper the more uncertain the activities, due to internal noise. Furthermore, the predicted distribution of outputs shows that the output distribution is less weighted at uncertain activities with large "error bars", and can lead to bimodal distributions with may only transmit one bit of information (two possible states about the environment).

# Further reading

(1) 'Modeling the chemotactic response of Escherichia coli to time-varying stimuli' by Yuhai Tu, Thomas S Shimizu, and Howard C Berg, *Proc. Natl. Acad. Sci. USA* 105: 14855–60 (2008).

(2) 'A simple coding procedure enhances a neuron's information capacity' by Simon Laughlin, Z. *Naturforsch.* 36: 910–12 (1981).

(3) 'Engineering aspects of enzymatic signal transduction: photoreceptors in the retina', by Peter B. Detwiler, Sharad Ramanathan, Anirvan Sengupta, and Boris I. Shraiman, *Biophys. J.* 79: 2801–17 (2000).

(4) 'Information flow and optimization in transcriptional regulation', by Gasper Tkacik, Curtis G. Callan Jr, and William Bialek, *Proc. Natl. Acad. Sci. USA* 105: 12265–270 (2008).

(5) 'Fold-change detection and scalar symmetry of sensory input fields', by Oren Shoval, Lea Goentoro, Yuval Hart, Avi Mayo, Eduardo Sontag, and Uri Alon, *Proc. Natl. Acad. Sci. USA* 107: 15995–16000 (2010).

(6) 'Response rescaling in bacterial chemotaxis', by Milena D. Lazova, Tanvir Ahmed, Domenico Bellomo, Roman Stocker, Thomas S. Shimizu, *Proc. Natl. Acad. Sci. USA* 108: 13870–5 (2011).

(7) 'Receptor Sensitivity in bacterial chemotaxis', by Victor Sourjik and Howard C. Berg, *Proc. Natl. Acad. Sci.* USA 99: 123–127 (2002).

(8) 'Chemotaxis: how bacteria use memory', by Nikita Vladimirov and Victor Sourjik, *Biol. Chem.* 390: 1097–1104 (2009).

# 12

# Beyond *E. coli* chemotaxis

In this book we chose the experimentally well-characterized chemotaxis pathway in the bacterium *E. coli* to introduce physical principles to sensing and signaling. We saw, using specific examples, the contributions which mathematics and physics have made to this sufficiently complex biological system. Topics we discussed are the role of receptor clustering in signal amplification, integration, and adaptation, the importance of robustness, the physical limits to the accuracy of sensing due to molecular diffusion, and reliable signaling in terms of information processing. These topics are, 'however', not only important for bacterial chemotaxis, but also for other biological systems. In addition to asking "how" things work, we have substantially moved towards the "why" questions. However, proper answers (to the "why" questions) would require a more holistic analysis of the bacterium with its energy budget and the cellular communities in which it lives, as well as evolutionary considerations.

The choice of topics was driven by personal favoritism, and was by no means all-inclusive. The emphasis was on topics related to sensing and signaling in the chemotaxis pathway, and a number of other equally interesting topics were unfortunately omitted. These include the workings of the flagellated rotary motor, different types and arrangements of flagella in different bacterial species, the cells' swimming behavior in a microscopic low Reynolds number environment so different from our macroscopic world, the various search strategies for finding food, and other bacterial motility mechanisms. Another aspect left out is the robustness of signaling pathways to changes in temperature. This robustness to temperature is somewhat surprising, as individual reaction rates are expected to be highly sensitive to temperature by Arrhenius' law.

Where is the field heading? Starting from what we know about *E. coli* chemotaxis, we can expand into other areas by softening some of the assumptions we made about the modularity and isolation of the pathway from other cellular processes. This would allow us to enter the realm of signaling cross-talk and signal integration with other signaling, metabolic, and gene-regulatory pathways. A particularly well-characterized area is the uptake of sugars by cells. While chemotaxis receptors deal with sensing, they are not directly responsible for the uptake of ligand molecules. Hence, how uptake of nutrients is coupled with chemotaxis is a natural extension of the bacterial chemotaxis work. Yet another important class of pathways is quorum-sensing, which allow bacteria to sense cell density, even of other bacterial species. These pathways are key in virulence and symbiosis with a host organism. In addition to bacterial chemotaxis and quorum-sensing, there are literally hundreds of other pathways in a cell, which are based on similar pathway architectures (so-called two-component systems, as they consist at their core of a sensor protein, as well as a response regulator protein). The big question concerns how signals are integrated or how interference is avoided.

How does knowledge in *E. coli* chemotaxis translate into other bacterial species? There are two relatively well characterized bacteria worth mentioning. *Rhodobacter sphaeroides* is a metabolically diverse bacterium, which has in addition to a membrane-bound cluster a second, cell-internal cluster, presumably to monitor the cell's metabolic state. The bacterium *Bacillus subtilis* is fascinating since it does not only have one adaptation mechanism, but three. Two of these are not based on receptor methylation. Open questions are how these alternative pathway designs reflect the lifestyle of the bacteria. However, the answers to the above listed questions are somewhat difficult to produce, since these pathways are much less understood than chemotaxis in *E. coli* and, hence, are less accessible to quantitative modeling. Despite significant differences between pathways among different species, all at least contain the same core proteins.

In the preceeding chapters we used a variety of quantitative methods from scientific disciplines other than traditional biology. In addition to the frequently discussed topics of diffusion and kinetic reaction laws, these methods include Boltzmann statistics, material properties of soft bio-molecules, control and information theory, and ways of analyzing data. Surprisingly, we rediscovered (or at least applied) concepts from engineering (such as integral feedback control) and telecommunications (mutual information). Where do these analogies come from? Presumably, problems bacteria have to solve are not that different from problems humans, and both types of organism are constrained by including energy availability, and the physical environment with certain statistical properties and fundamental laws.

It is worth saying a few words about the choices we made for modeling. The process of understanding biological complexity certainly requires some rather drastic measures of simplification. If we did not make these, we could equally conduct molecular-dynamics simulations of everything in a cell. However, even if that were feasible, such a model would be nearly as complex as the real cell, and we would learn relatively little.

The choices of what to include in a model appear at first sight somewhat arbitrary and more art than science. What to include, however, really depends on the level of detail for describing a biological process. An important guiding principle is the degree at which the process of interest is isolated from the many other processes which constantly occur inside a cell. This would justify modeling the pathway independently. This isolation (or modularity) can be in time, i.e. that other processes are either much faster and hence are averaged out, or are much slower so that they effectively do not change. For instance, in the description of signaling we left out processes such as gene regulation and potential metabolic changes, as they are much slower. We also left out very fast processes at the molecular or atomistic level, such as diffusion in the cytosol or subtle conformational changes in protein conformations, except for the motor, whose switching is rather slow. For instance, the receptor activity was treated as the average probability to be in the active conformation. Alternatively, the isolation can be in space, so that processes do not physically interact at all. This is to some degree satisfied here, as most of the signaling occurs at the receptor cluster.

Of course, some choices can lead to wrong results—but even that need not be a major problem. Modeling is really about hypothesis generation and guiding new experiments, which are most effective in unraveling the biological truth. Even wrong models can fulfil this purpose, but need to be revised in light of new conflicting data. Numerous classical examples can be taken from astronomy and cosmology, but also from the microscopic world. When diffusion was initially described by Einstein, he only considered the position of the particle as a random variable but neglected its momentum. Although he successfully derived the distribution of diffusing particles in time, the neglect of the momenta was a rather severe simplification (also called *contraction*) and led to the unphysical result that at short times particles had an infinite velocity. Later, less contracted models were able to remedy this problem. Overall, modeling, mathematics, and physics can be considered tools to see beyond the data in order to find trends and helpful interpretations. In this sense, mathematics has been described as the "new microscope".

The big motivator for such modeling is the search for general principles in biological systems, which might ultimately be either engineering principles as biological systems try to solve problems having to do with "improving life". Example are the use of filters, amplifiers, and feedback mechanisms by biological systems to properly process and interpret information. Alternatively, these principles might be physical as physics is the most fundamental description of nature. An example is the fundamental physical limit in accuracy of sensing chemicals as set by ligand diffusion. While the determination of such general principles is a tall order, an intermittent way forward is to very carefully study particular well-characterized systems and to identify, based on explicit examples, more and more general principles. To be general, these should be independent of many molecular details and cell types. Again, integral feedback control as a way to track a particular desired cellular property in time does not only appear in bacterial chemotaxis, but was also found to be important in embryonic development and homeostasis. Similarly, fold-change detection is not only relevant to bacterial chemotaxis but also for robust gene expression.

While mathematical modeling is really nothing new in the field of biology, the explosion of high-quality and quantity data has created a real need for mathematical analysis and prediction. It is not a "luxury" to spend time on modeling any more, but the vast amounts of data have little meaning if not seen through the lens of the "mathematical microscope". In contrast to the distant past, the scientific field of the last fifty years has been stuck in endless specialization, driven by the emergence of new measurement technologies. However, to solve the big questions in biology, the integration of the various scientific disciplines will be the fruitful way forward. In a sense, we return to the original path of science.

## Chapter summary

- The well-characterized *E. coli* chemotaxis pathway can be considered a modeling playground to go from "how" to "why" questions. In the future the bacterial chemotaxis pathway, here considered in isolation, will probably be considered in the context of other pathways in the cell, and in other, more complicated bacteria.

- The topics covered indicate a strong similarity between how we solve problems in our macroscopic world and how bacteria (or cells in general) solve problems in the microscopic world.

- Modeling is about making good simplifying assumptions, isolating aspects of large, complicated biological systems for quantitative understanding. This isolation can sometimes be justified due to separation of important dynamics in time or space.

- The application of models to biology is really about producing hypotheses and guiding new experiments in fruitful directions. Even "wrong" models can fulfil this purpose, but need to be revised in light of conflicting data.

- Another necessity of modeling, now more than decades ago, is its role in summarizing and explaining large amounts of complicated data, not being comprehensible otherwise. In a sense, modeling can be considered the "new microscope".

- The big picture of the presented research is about finding general principles in biological systems, and extending the success of physics from non-living to living matter.

## Further reading

(1) 'Quantification of Behavior Sackler Colloquium: Searching for simplicity in the analysis of neurons and behavior', by Greg J. Stephens, Leslie C. Osborne C, and William Bialek, *Proc. Natl. Acad. Sci. USA* 108 Suppl 3:15565–71 (2011)

(2) 'Transcription by the numbers redux: experiments and calculations that surprise', by Hernan G. Garcia, Alvaro Sanchez, Thomas Kuhlman, Jane Kondev, and Rob Phillips, *Trends Cell. Biol.* 20: 723–733 (2010).

(3) 'Mathematics is biology's next microscope, only better; biology is mathematics' next physics, only better', by Joel E. Cohen, *PLoS Biol.* 2: e439 (2004).

(4) 'Back to the future: education for systems-level biologists', by Ned S. Wingreen and David Botstein, *Nat. Rev. Mol. Cell Biol.* 7: 829–832 (2006).

(5) *Physical Biology of the Cell*, by Rob Phillips, Jane Kondev, and Julie Theriot (Garland Science, 2008).

(6) *Statistical Thermodynamics of Nonequilibrium processes*, by Joel Keizer (Springer Press, 1987).

(7) 'Dynamic assignment and maintenance of positional identity in the ventral neural tube by the morphogen sonic hedgehog', by Eric Dessaud *et al.*, *PLoS Biol.* 8: e1000382 (2010).

(8) 'Making an impact in biology', by Robert G. Endres, *Phys. World* 23: 16–17 (2010).

(9) 'Thermal robustness of signaling in bacterial chemotaxis', by Olga Oleksiuk, Vladimir Jakovljevic, Nikita Vladimirov, Ricardo Carvalho, Eli Paster, William S. Ryu, Yigal Meir, Ned S. Wingreen, Markus Kollmann, Victor Sourjik, *Cell* 145: 312–21 (2011).

(10) 'The three adaptation systems of *Bacillus subtilis* chemotaxis', by Christopher V. Rao, George D. Glekas, and George W. Ordal, *Trends Microbiol.* 16: 480–7 (2008).

(11) 'Spatial organization in bacterial chemotaxis', by Victor Sourjik, and Judith P. Armitage, *EMBO J.* 29: 2724–2733 (2010).

(12) 'The switching dynamics of the bacterial flagellar motor', by Siebe B van Albada, Sorin Ta brevenase-Nicola, and Pieter Rein ten Wolde, *Mol. Syst. Biol.* 5: 316 (2009).

(13) 'Life at low Reynolds number', by Edward M. Purcell, *Am. J. Phys.* 45: 3–11. (1977).

(14) 'Bacterial motility: links to the environment and a driving force for microbial physics', by James G. Mitchell, and Kazuhiro Kogure, *FEMS Microbiol. Ecol.* 55: 3–16 (2006).

# More techniques

# Appendix

In this appendix we provide derivations and describe additional techniques from physics and mathematics. All these are used, in one way or another, in this book.

## A.1 Derivation of the fluctuation–dissipation theorem

Due to its fundamental importance, we sketch the derivation of the FDT briefly. The average value of our observable can be written in terms of the probability distribution $W(x,0)$ to find protein at position $x$ at time zero (before the perturbation was switched on) and the transition probabilities $P(x',t|x,0)$ to go from $x$ at $t=0$ to $x'$ at $t'$ via

$$\langle x(t) \rangle = \int dx' \int dx\, x'\, P(x',t|x,0)\, W(x,0). \qquad (A.1)$$

The probability distribution $W(x,0)$ is an equilibrium distribution, and hence can be obtained from the Boltzmann distribution using the energy $E(x)$ of the system (note the integral representation)

$$W(x,0) = \frac{e^{-\frac{E(x)}{k_B T}}}{\int dx' e^{-\frac{E(x')}{k_B T}}}. \qquad (A.2)$$

For small perturbations $f_0/(k_B T)$ we can expand the right-hand side and obtain

$$W(x,0) \approx W_0(x)\left(1 - \frac{f_0 x}{k_B T}\right), \qquad (A.3)$$

where we set from now on $\langle x \rangle_0 = 0$ for convenience. This can be used to rewrite the average value of the observable

$$\langle x(t) \rangle = -\frac{f_0}{k_B T} A(t) = f_0 \int_{-\infty}^{t} \theta(\tau)\chi(t-\tau)d\tau \qquad (A.4)$$

with the *auto-correlation function*[1] $A(t) = \langle x(t)x(0) \rangle$ and the linear response result from Eq. (3.35). To get rid off the integral, we differentiate with respect to time using the *Leibniz rule*[2] and obtain

$$\chi(t) = \frac{1}{k_B T} \frac{d}{dt}[A(t)\theta(t)]. \qquad (A.5)$$

[1] Autocorrelation is the cross-correlation of a signal with itself, often expressing the similarity between two observations as a function of their time separation.

[2] The Leibniz rule allows the calculation of derivatives of integrals by $\frac{d}{d\alpha}\int_{a(\alpha)}^{b(\alpha)} f(x,\alpha)dx = \frac{db(\alpha)}{d\alpha}f(b(\alpha),\alpha) - \frac{da(\alpha)}{d\alpha}f(a(\alpha),\alpha) + \int_{a(\alpha)}^{b(\alpha)} \frac{\partial}{\partial\alpha}f(x,\alpha)dx$.

The *Heaviside step function*[3] $\theta(t)$ ensures that $\chi(t)$ is zero for negative times when there is no perturbation. To eliminate the time derivative, we plug in the Fourier transforms for the time-dependent functions, and thus obtain a factor $i\omega$ on the right-hand side. Note that the product becomes a convolution in Fourier space. We do not need all the technical details here if we remember that the Fourier transform of a symmetric real function is real, while the Fourier transform of an asymmetric function is purely imaginary. Introducing the *sign* function $\mathrm{sgn}(t)$,[4] we can write

$$A(t)\theta(t) = \underbrace{A(t)}_{\text{symmetric}} + \underbrace{A(t) \cdot \mathrm{sign}(t)}_{\text{asymmetric}}. \tag{A.6}$$

We now use the *Wiener–Khinchin theorem*[5] $\hat{S}(\omega) = 2\hat{A}(\omega)$, i.e. that the power spectrum of a stationary process (in absence of the perturbation) is twice the Fourier transform of the auto-correlation function. As a result we obtain an expression for the response function

$$\hat{\chi}(\omega) = \frac{i\omega}{k_B T}\left[\frac{\hat{S}(\omega)}{2} + \text{imaginary stuff}\right]. \tag{A.7}$$

To finally obtain the power spectrum, we simply take the imaginary part (the first part upon multiplication with prefactor) of $\hat{\chi}(\omega)$, which proves the FDT.

## A.2   Variational principles and the Euler–Lagrange equation

Although *variational principles* can be regarded as purely mathematical tools for solving optimization problems, they have immense importance in all kinds of areas of physics and are used several times in this book, e.g. in deriving the Boltzmann distribution (Chapter 3.2). Here we will introduce another important example, the *Euler–Lagrange equation*.

The Euler–Lagrange equation is the solution to finding extrema of the functional[6] in the form of

$$S[y(t)] = \int_{x_1}^{x_2} L(x, y, y')dx, \tag{A.8}$$

where the solution is generally a second-order differential equation. In physics, the meaning of $S$ is the *action*[7] which is extremal due to *Hamilton's principle* with $x$ the time. Integrand $L$ is the *Lagrangian*, which contains all the important information about a physical system. In classical mechanics, this approach can be considered the integral version of *Newton's laws of motion*. However, it is also similar to the familiar problem of finding the local maxima or minima of a function by calculating the locations where the function's derivative is zero. The formalism can be used to find optimal solutions of a variety of mathematical problems.

For instance, the function $L$ could be the arc length between two points, and the optimal (shortest) length is a straight line

To derive the Euler–Lagrange equation we calculate the variation of $S$

$$\delta S = \frac{\partial}{\partial h} S(y + h\Delta y)|_{h=0} \qquad (A.9)$$

with $\Delta y(x)$ an arbitrary differential function, which vanishes at the boundary, $\Delta y(x_1) = \Delta y(x_2) = 0$. To find the extrema, the variation must be zero, i.e. $\delta S = 0$, or

$$\frac{\partial}{\partial h} \int_{x_1}^{x_2} L(x, y + h\Delta y, y' + h\Delta y')dx|_{h=0} = 0. \qquad (A.10)$$

Using the *chain rule*[8] for differentiation of a function of multiple variables, Eq. (A.10) simplifies to

$$\int_{x_1}^{x_2} \left[ \frac{\partial L}{\partial y} \cdot \frac{d(y + h\Delta y)}{dh} + \frac{\partial L}{\partial y'} \cdot \frac{d(y' + h\Delta y')}{dh} \right] dx \qquad (A.11)$$

$$= \int_{x_1}^{x_2} \left( \frac{\partial L}{\partial y} \Delta y + \frac{\partial L}{\partial y'} \Delta y' \right) dx \qquad (A.12)$$

$$= \int_{x_1}^{x_2} \Delta y \frac{\partial L}{\partial y} dx + \left[ \Delta y \frac{\partial L}{\partial y'} \right] - \int_{x_1}^{x_2} \Delta y \frac{\partial}{\partial x} \left( \frac{\partial L}{\partial y} \right) dx, \qquad (A.13)$$

where we used the *integration-by-parts rule*[9] to reach the last line. However, $\Delta y(x_1) = \Delta y(x_2) = 0$, so that the middle term in Eq. (A.13) is zero,

$$\int_{x_1}^{x_2} \Delta y \left[ \frac{\partial L}{\partial y} dx - \frac{\partial}{\partial x} \left( \frac{\partial L}{\partial y'} \right) \right] dx = 0. \qquad (A.14)$$

For arbitrary $\Delta y$, the integral can only be zero if the term in [...] vanishes (*fundamental lemma of calculus of variations*) and we obtain the Euler–Lagrange equation

$$\frac{\partial L}{\partial y} - \frac{\partial}{\partial x} \left( \frac{\partial L}{\partial y'} \right) = 0. \qquad (A.15)$$

[8]Chain rule for differentiation: $\frac{d}{dx} f[g(x)] = \frac{df}{dg} \cdot \frac{dg}{dx}$.

[9]Integration-by-parts rule: $\int f(x) \frac{dg}{dx} dx = f(x)g(x) - \int \frac{df}{dx} g(x)dx$.

## A.3   Gillespie simulations

A Gillespie simulation provides the same probability distribution as the corresponding master equation and hence is exact. However, many simulations might be required to provide good statistics. In addition, there are other implementations of kinetic Monte Carlo simulations, aiming for increased speed at the expense of accuracy.

The basic Gillespie algorithm is easily implemented. Suppose there are different chemical species with certain copy numbers, as well as $i = 1, 2, \ldots$ possible chemical reactions with rates $r_i$ at a particular point in time. The total rate that something happens is $\Gamma = \sum_{i=1} r_i$. Running a Gillespie simulation on this system requires repeated drawing of two random numbers using the following iterative scheme.

**Step 1:** Determine the next occurring reaction by drawing random number $R_1$ between 0 and 1, and choosing reaction $i$ accordingly. Specifically, choose reaction $i = 1$ if $0 \leq R_1 < r_1/\Gamma$, choose reaction $i = 2$ if $r_1/\Gamma \leq R_1 < (r_1 + r_2)/\Gamma$, and so on.

**Step 2:** Update copy molecule numbers according to reaction stoichiometry.

**Step 3:** Increment simulation time by $\tau = -\ln(R_2)/\Gamma$, where $R_2$ is a second random number evenly distributed between 0 and 1. This time increment $\tau$ is obtained by inverting the waiting-time distribution $R_2 = e^{-\Gamma\tau}$ (see Fig. A.1 for a graphical derivation). Then go to step 1 and iterate.

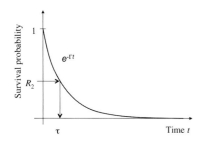

**Fig. A.1** Graphical derivation of the time increment $\tau$ used in the Gillespie algorithm. Random number $R_2$ is chosen between 0 and 1, and the corresponding time increment can be read off on the horizontal axis.

## A.4    Fokker–Planck approximation

For large systems, generally the master equation can not be solved and Gillespie simulations become too time consuming. Here we describe an approximate equation, the Fokker–Planck equation, valid when molecule numbers are large. In this limit, the molecule numbers can be approximated as continuous variables and relevant functions can be expanded around points of interest.

Consider our exemplar master equation (4.21), which is reproduced here for clarity reasons

$$\frac{dp_N}{dt} = -(f_N + g_N)p_N + f_{N-1}p_{N-1} + g_{N+1}p_{N+1}. \tag{A.16}$$

In the continuum limit, we replace $p_N$, $f_N$, and $g_N$ by functions $p(N)$, $f(N)$, and $g(N)$. Function products $f(N-1)p(N-1)$ and $g(N+1)$ $p(N+1)$, simply termed $F(N)$ for short, can then be expanded to second order

$$F(N \pm 1) \approx F(N) \pm \frac{\partial F(N)}{\partial N} + \frac{1}{2}\frac{\partial^2 F(N)}{\partial N^2}. \tag{A.17}$$

When plugged into the master equation, Eq. (A.16), we obtain the Fokker–Planck equation

$$\frac{\partial p(N)}{\partial t} = -\frac{\partial}{\partial N}[v(N)p(N)] + \frac{1}{2}\frac{\partial^2}{\partial N^2}[D(N)p(N)]. \tag{A.18}$$

This equation is quite intuitive: $v(N) = f(N) - g(N)$ is an effective drift velocity, describing the deterministic behavior (like in response to a force), and $D(N) = f(N) + g(N)$ is an effective diffusion constant, which describes the noise (like a temperature). The Fokker–Planck equation is really a continuity equation

$$\frac{\partial p(N)}{\partial t} = -\frac{\partial j(N)}{\partial N}, \tag{A.19}$$

where $j(N) = v(N)p(N) - 1/2 \cdot \partial[D(N)p(N)]/\partial N$ is the probability current. This comes in handy for obtaining steady-state solutions.

While the time-dependent problem is generally not solvable, the similarity of the Fokker–Planck equation and the Schrödinger equation from quantum mechanics allows the application of a number of useful techniques such as operators or path integrals.

## A.5   Derivation of the Langevin noise

We introduced the Langevin small noise approximation in Section 4.6. In that chapter we derived the noise terms heuristically. However, the noise terms can also be derived systematically from the underlying master equation for the molecule copy number $N = 0, 1, 2, \ldots$. We start by restating exampler master Eq. (4.21)

$$\frac{\partial p_N}{\partial t} = \tilde{k}V p_{N-1} + \gamma(N+1)p_{N+1}$$
$$- \tilde{k}V p_N - \gamma N p_N, \qquad (A.20)$$

where we wrote $\tilde{k}V$ instead of $k$ to introduce the reaction volume $V$.

We expect $p_N$ to peak around the macroscopic value $Vn(t)$ (where $n$ is now a concentration) with a width of order $N^{1/2} \sim V^{1/2}$. To use this assumption, we now define the following separation of $N$ into macroscopic and fluctuating parts via

$$N(t) = Vn(t) + V^{1/2}\eta(t), \qquad (A.21)$$

allowing us to expand terms using $V$ as a large parameter.[10] We also define the step operators

$$\mathbf{E}^{+1}F(N) = F(N+1) \qquad (A.22)$$
$$\mathbf{E}^{-1}F(N) = F(N-1) \qquad (A.23)$$

for any arbitrary continuous function $F(\ldots)$. Using Eqs. (A.21), in the limit of large $V$ the step operators adopt the differential form

$$\mathbf{E}^{\pm 1} = 1 \pm V^{-1/2}\frac{\partial}{\partial \eta} + \frac{1}{2}V^{-1}\frac{\partial^2}{\partial \eta^2} \pm \ldots \qquad (A.24)$$

with higher-order terms neglected. Transforming from the old variable $N$ to the new small variable $\eta$, we obtain the relations

$$p_N(t) \to \Pi(\eta, t) \qquad (A.25)$$
$$V^{1/2}\frac{\partial}{\partial N}p_N(t) = \frac{\partial}{\partial \eta}\Pi(\eta, t). \qquad (A.26)$$

With the above relations we transform the master equation, now written in step operators,

$$\frac{\partial p_N}{\partial t} = \tilde{k}V(\mathbf{E}^{-1} - 1)p_N + \gamma(\mathbf{E}^{+1} - 1)N p_N$$
$$- (\tilde{k}V + \gamma N)p_N \qquad (A.27)$$

[10] This derivation is called the van Kampen expansion.

into

$$\frac{\partial \Pi}{\partial t} - V^{1/2}\frac{dn}{dt}\frac{\partial \Pi}{\partial \eta}$$

$$= \tilde{k}V^{1/2}\left[-\frac{\partial}{\partial \eta} + \frac{1}{2}V^{-1/2}\frac{\partial^2}{\partial \eta^2}\right]\Pi(\eta, t)$$

$$+ \gamma V^{1/2}\left[\frac{\partial}{\partial \eta} + \frac{1}{2}V^{-1/2}\frac{\partial^2}{\partial \eta^2}\right](n + V^{-1/2}\eta)\Pi(\eta, t). \tag{A.28}$$

The terms proportional to $V^{1/2}$ produce the macroscopic equation $dn/dt$ (Eq. (4.1)). In contrast, the terms proportional to $V^0$ produce

$$\frac{\partial \Pi}{\partial t} = \frac{1}{2}(\tilde{k} + \gamma n)\frac{\partial^2 \Pi}{\partial \eta^2} + \gamma\frac{\partial}{\partial \eta}(\eta \Pi). \tag{A.29}$$

Importantly, from Eq. (A.29) it is possible to derive equations for the average and variance of the fluctuations. Multiplication of Eq. (A.29) by $\eta$ and $\eta^2$ with subsequent integration yields, respectively,

$$\frac{\partial \langle \eta \rangle}{\partial t} = -\gamma\langle \eta^2 \rangle \tag{A.30}$$

$$\frac{\partial \langle \eta^2 \rangle}{\partial t} = \tilde{k} + \gamma n - 2\gamma\langle \eta^2 \rangle. \tag{A.31}$$

At steady state, we finally obtain

$$\langle \eta \rangle_s = 0 \tag{A.32}$$

$$\langle \eta^2 \rangle_s = \frac{\tilde{k}}{\gamma}. \tag{A.33}$$

If we stop the expansion in $V$ at the second moment, we assume that fluctuations are Gaussian distributed. After introducing Fourier transformations in the next section of this chapter, we will show that the Master equation and the Langevin approximation indeed produce identical Gaussian noises.

## A.6    Time versus frequency domain

Often processes in biology are time dependent (at least the interesting ones). However, sometimes it is advantageous to calculate properties not in the time domain, but in the frequency domain. This means that one decomposes a time-dependent process into a sum or integral over frequency components. This is analogous to describing a musical chord in terms of the notes being played.

In mathematics, this decomposition is achieved by the Fourier transform, which defines a relationship between a signal in the time domain and its representation in the frequency domain. The Fourier transform of $x(t)$ is defined by

$$X(f) = \int_{-\infty}^{\infty} x(t)e^{-i2\pi ft}dt \tag{A.34}$$

with its inverse transform given by

$$x(t) = \int_{-\infty}^{\infty} X(f)e^{i2\pi tf}df. \qquad (A.35)$$

Here, frequency $f$ is measured in Hertz (Hz= $s^{-1}$). Alternatively, one can use angular frequency $\omega = 2\pi f$, which would produce frequency in units of radians per second and introduce a factor $1/(2\pi)$ in Eq. (A.35).[11]

Being a transform, no information is created or lost in the process, therefore the original signal can always be recovered given the Fourier transformation. The Fourier transform of a time-dependent signal is a real or complex valued function of the frequency. Complex function values or numbers can either be represented by their real and imaginary parts ($\mathrm{Re}(X)$ and $\mathrm{Im}(X)$ respectively), or by their magnitude and phase ($|X|$ and $\Phi$ respectively). Mathematically, switching between the two representations for a given complex value can be done by

$$|X| = \sqrt{\mathrm{Re}(X)^2 + \mathrm{Im}(X)^2} \quad \text{and} \quad \Phi = \tan^{-1}\left(\frac{\mathrm{Im}(X)}{\mathrm{Re}(X)}\right), \qquad (A.36)$$

or equivalently, by

$$\mathrm{Re}(X) = |X|\cos(\Phi) \quad \text{and} \quad \mathrm{Im}(X) = |X|\sin(\Phi). \qquad (A.37)$$

For illustration we Fourier-transform a 4-ms unit pulse shown in Fig. A.2. In the frequency domain we obtain a continuous signal with a maximum at 0 Hz and decaying oscillations with a well-defined period to either side. Specifically, the magnitude is zero at multiples of 250 Hz. If we would plot the Fourier transform of a unit pulse of 8 ms instead, we would get zeros at multiples of 125 Hz. This shows that the zeros are reciprocals of the pulse duration, indicating an inverse relationship between time and frequency. Generally, longer time spans relate to smaller frequency spans. This trade-off is called the *uncertainty principle*.[12]

The Fourier transform of the unit pulse of duration $\tau$ can easily be calculated analytically:

$$X(f) = \int_{-\tau/2}^{\tau/2} e^{-i2\pi ft}dt = \left[\frac{i}{2\pi f}e^{-2\pi ft}\right]_{-\tau/2}^{\tau/2} = \tau \cdot \frac{\sin(\pi\tau f)}{\pi\tau f}. \qquad (A.38)$$

Hence, a change in duration changes the magnitude of the Fourier transform, as well as the spacing between the zeros in the sinus function.

Let us now turn to even simpler signals. Sine and cosine waves are signals that by definition contain only one frequency. For instance, calculating the Fourier transform of $x(t) = \cos(2\pi f_0 t)$ produces $X(f) = 1/2[\delta(f - f_0) + \delta(f + f_0)]$. Adding a delay $\Delta\tau$ results in $X(f) = 1/2[\delta(f - f_0) + \delta(f + f_0)]e^{-i2\pi f\Delta\tau}$. Specifically, for $\Delta\tau = 1/(4f_0)$ the signal becomes $x(t) = \sin(2\pi f_0 t)$ and the corresponding Fourier transform $X(f) = i/2[\delta(f + f_0) + \delta(f - f_0)]$ is purely imaginary.[13] As a result, adding a delay is equivalent to multiplying the Fourier transform by

[11] We will use the latter definition in the remainder of the book. Yet another popular convention is to "split" the factor $2\pi$ evenly between the Fourier transform and its inverse.

[12] This relates to Heisenberg's uncertainty principle from quantum physics, which states that certain pairs of physical properties of a particle, such as position and momentum, cannot be simultaneously measured with arbitrary precision.

[13] Here we used the Euler formula $e^{i\phi} = \cos\phi + i\sin\phi$.

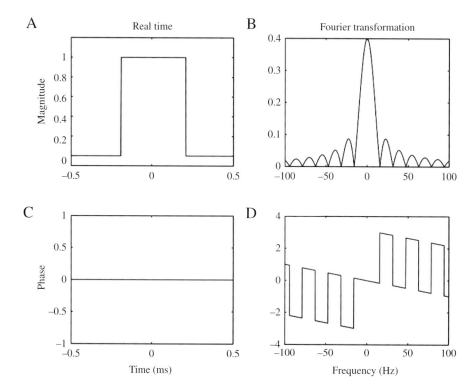

**Fig. A.2** Example of a Fourier transformation. (A) Magnitude and (C) phase of original pulse, a 0.4ms unit pulse. As pulse is real, the phase is zero. (B) Magnitude and (D) phase of transformation into frequency domain.

a phase factor. However, the magnitude stays the same. Alternatively, using the representation by real and imaginary parts, transforms that were purely real without a delay can become purely imaginary with a delay. Analogously, shifting the Fourier transform (a delay in frequency space) introduces a phase in the time domain.

To illustrate the usefulness of the Fourier transform, we briefly look at two examples of high general importance: linear response theory and the Langevin equation.

The linear response $R(t)$ of a system from equilibrium due to a thermodynamically conjugate force $F(t)$ defines the generalized susceptibility or response function $\chi(t)$ via

$$R(t) = \int_{-\infty}^{t} \chi(t - t')F(t')dt', \qquad (A.39)$$

where we have taken the average $\langle R(t)\rangle = 0$. The generalized susceptibility depends on the properties of the system and completely characterizes its response to small perturbations. Effectively, in Eq. (A.39) the value of $R(t)$ does not only depend on the present value of $F(t)$, but also on past values. In other words, $R(t)$ is a weighted sum of the previous values of $F(t')$ with the weights given by the generalized susceptibility $\chi(t - t')$.

Importantly, while Eq. (A.39) represents a convolution of two functions $F$ and $\chi$, inverse Fourier transforming leads to the unconvoluted simple product

$$\hat{R}(\omega) = \hat{\chi}(\omega)\,\hat{F}(\omega) \tag{A.40}$$

with angular frequency $\omega = 2\pi f$. This allows solving for the generalized susceptibility in terms of stimulus and response. Specifically, if we consider the impulse response, i.e. that $F(t)$ represents a sharp pulse in time, its Fourier transform is a constant so that $\hat{R}$ is just the generalized susceptibility.

We now solve the Langevin equation (4.33), analytically to demonstrate that it produces the same variance in the molecule concentration as the expansion of the master equation. To describe fluctuations around steady-state, we introduce the new variable $\delta n(t) = n(t) - \bar{n}$, where $\bar{n}$ is the steady-state copy number. Plugging this into Eq. (4.33) leads to the linearized equation

$$\frac{d(\delta n)}{dt} = -\gamma \delta n + \eta. \tag{A.41}$$

To solve this equation we use the inverse Fourier transform $\delta n(t) = \int_{-\infty}^{\infty} \delta \hat{n}(\omega) e^{i\omega t} dt$. On the left-hand side of Eq. (A.41), the time derivative produces a factor $\omega$, resulting in

$$(\gamma - i\omega)\delta\hat{n} = \hat{\eta}. \tag{A.42}$$

Now we calculate the power spectrum $S_n(\omega)$ of the fluctuations, defined by

$$S_n(\omega) = \langle \delta\hat{n}(\omega)\delta\hat{n}^*(\omega) \rangle = \langle |\delta\hat{n}|^2 \rangle = \frac{\langle |\hat{\eta}|^2 \rangle}{\gamma^2 + \omega^2} = \frac{2k}{\gamma^2 + \omega^2} \tag{A.43}$$

with superscript $*$ denoting the complex conjugate and the noise strength derived in Section 4.6. To move from the frequency-dependence of the fluctuations to the total variance we integrate over all frequencies

$$\langle (\delta n)^2 \rangle = \int_{-\infty}^{\infty} \frac{d\omega}{2\pi} \langle |\delta\hat{n}|^2 \rangle = \frac{k}{\gamma}. \tag{A.44}$$

As you can see, upon division by volume $V$ this result is identical to Eq. (A.33) from the master equation.

## A.7   Model fitting to data

Often we want to represent complicated data using a simple model of compact mathematical equations. This provides a condensed way of summarizing the data, along with valuable parameters describing the biological process such as rate constants. For instance, take the formula for the exponential decay of a protein

$$N(t) = N_0 \cdot e^{-\lambda t} \tag{A.45}$$

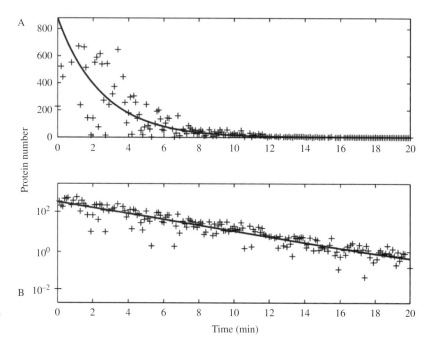

**Fig. A.3** Fitting example using decaying protein. Non-linear (A) and linear (B) regression are compared. Linearization of plot in (A) leads to a skewing of the Gaussian scatter in (B).

[14] The inverse lifetime $\lambda$ is related to the half-life $\tau_{0.5}$, the time to reduce the number by a factor 2, via $\lambda = \ln 2/\tau_{0.5}$.

[15] Here one maximizes the logarithm of the likelihood function, $P(\text{data}|\text{parameters})$, i.e. the conditional probability of the data given the model parameters.

with time $t$ as the independent variable and parameters $N_0$ and $\lambda$ representing the initial copy number of proteins and the decay constant, respectively.[14] Fitting this model to the data would provide either the parameters which fit the data best (best-fitting parameters) or allow the comparison with other models to find the best model (model selection). For the latter, be aware of introducing dependent parameters: for instance, in $N(t) = N_0 \exp(-\lambda t + K)$ parameters $N_0$ and $K$ are dependent and must not both be used. To perform a fit, we need a measure of closeness between the data and the model. Often one uses the $\chi^2$ goodness of fitness function

$$\chi^2(N_0, \lambda) = \sum_{i=1}^{N} \left( \frac{N_i - N_{N_0, \lambda}(t)}{\sigma_i^2} \right)^2, \qquad (A.46)$$

which can easily be derived from an approach called *maximum-likelihood estimation*[15] and the assumption that the data is Gaussian distributed. (The latter is expected from the central limit theorem.) Intuitively, using the squared error $(N_i - N_{N_0, \lambda}(t))^2$ between data $N_i$ and model $N_{N_0, \lambda}(t)$ instead of, say the absolute error $|N_i - N_{N_0, \lambda}|$, ensures that errors are small in general instead of having some very large and some very small errors. (Just the error $(N_i - N_{N_0, \lambda}(t))$ is useless, as multiple errors would be able to cancel themselves.) Minimizing this function with respect to parameters $N_0$ and $\lambda$ provides the best-fitting parameters. This can either be done analytically, such as for the simple case of *linear regression* discussed next, or numerically.

Before computers were invented, people tried to linearize models. In this case it is sufficient to fit a simple straight line to the data (linear regression). In our example this is achieved by taking the natural logarithm

$$\ln N = \ln N_0 - \lambda \cdot t \tag{A.47}$$

which can be fitted to a line by using $N_{a,b}(t) = a + b \cdot t$ in Eq. (A.46). Next we minimize this function by taking the derivatives with respect to the two parameters and by setting the derivatives to zero. Solving for $a$ and $b$ leads to the best-fitting parameters. From these, the original parameters $N_0$ and $\lambda$ can be obtained. However, even when the model can be linearized, the assumptions for applying linear regression correctly are often not satisfied, e.g. that the scatter of data around the line follows a Gaussian distribution. In fact, the logarithmic transformation Eq. (A.47) used to linearize Eq. (A.45) visibly skews the Gaussian scatter as shown in Fig. A.3.

To fit non-linear models to data, the method of *non-linear regression* can generally be applied. Now, Eq. (A.46) needs to be minimized numerically to obtain the parameter values $N_0$ and $\lambda$, which best describe the data. This minimization occurs in the 2D space of our two parameters, and needs to be done iteratively using initial guesses for the parameter values and then moving systematically in a favorable direction in this parameter space. A popular method for this purpose is the *Levenberg–Marquardt algorithm*.

However, we are not done yet. Finally, the fitting result needs to be interpreted and evaluated. First, parameter values need to be physically relevant. For instance, fitted rate constants must not be negative or of unreasonable size (e.g. years instead of minutes). Second, if different models are compared, one needs to ensure that the most suitable model is not only the best because it has more model parameters (overfitting) and hence more degrees of freedom to fit the data.

Overfitting can be avoided by adding a penalty term proportional to the number of model parameters to the $\chi^2$ function, derivable from an approach called *Bayesian inference*. Bayesian inference relies on Bayes' theorem

$$P(\text{parameters}|\text{data}) = \frac{P(\text{data}|\text{parameters})P(\text{parameters})}{P(\text{data})}, \tag{A.48}$$

which follows from the fact that the joint probability of observing two things, $a$ and $a$, is given by $P(a,b) = P(a|b)P(b) = P(b|a)P(a)$. Specifically, Eq. (A.48) means that the posterior probability $P(\text{parameters}|\text{data})$ of obtaining model parameters given the data equals the likelihood function $P(\text{data}|\text{parameters})$, the conditional probability of observing the data given the model, times the prior probability $P(\text{parameters})$ of observing the parameters. Finally, everything is divided by the probability of observing the data $P(\text{data})$. As the measurement is already finished, the latter can be considered a normalization constant.

Third, confidence intervals are needed for the fitted parameter values to describe how well the obtained estimates describe the data. Such confidence intervals can be obtained from the standard deviation of the posterior probability distributions obtained from Bayesian inference. Confidence intervals indicate that the true (unknown) parameter values are within the interval boundaries with high chance. If this interval length is narrow we deduced the true parameter values very precisely.

## A.8   Principal component analysis

Principal component analysis (PCA) involves a mathematical procedure that transforms a number of possibly correlated variables into a smaller number of linearly uncorrelated variables called principal components. All the principal component together account for the variability in the data, with the first principal component contributing the most, the second principal component the second most, and so on.

As an example, measured dose-response curves $r(d)$ may display substantial variability (see illustrative curves in Fig. A.4A). Despite the large variation (mainly in the amplitude), the (two) values of the sample data within a dose-response curve are highly correlated, e.g. have all either a relatively low or a relatively high value. These correlations weigh against the use of standard $\chi^2$ fitting, which is only appropriate if the errors are independent from data point to data point. To obtain a representation of the data where errors are independent, we calculate the principal components of the data.

In PCA we evaluate the covariance matrix, describing the correlations between the response measurements different stimuli. For the illustrative data in Fig. A.4 with only two stimuli, $d_1$ and $d_2$, the covariance matrix is given by

$$C = \frac{1}{M} \sum_{i=1}^{M} \boldsymbol{r}_i \boldsymbol{r}_i^T = \frac{1}{M} \sum_{i=1}^{M} \begin{pmatrix} r_i(d_1)r_i(d_1) & r_i(d_1)r_i(d_2) \\ r_i(d_1)r_i(d_2) & r_i(d_2)r_i(d_2) \end{pmatrix}$$
$$= \begin{pmatrix} \langle r(d_1)r(d_1) \rangle & \langle r(d_1)r(d_2) \rangle \\ \langle r(d_1)r(d_2) \rangle & \langle r(d_2)r(d_2) \rangle \end{pmatrix},$$

**Fig. A.4** Illustration of principal component analysis (PCA). (A) Five "dose-response curves" of sample data each with only two data values. Stimuli are indicated by arrows. (B) Scatter plot of the response with the new local coordinate system from the PCA. Redrawn from Ref. (8).

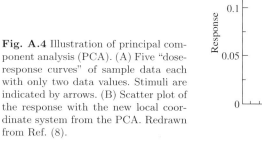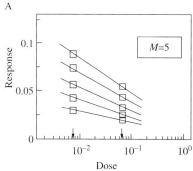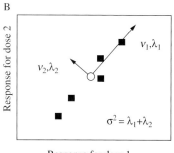

where superscript $T$ denotes the transpose and $\langle \ldots \rangle$ the average over the $M$ dose-response curves. Diagonalization of the covariance matrix provides the eigenvectors (here $\mathbf{v}_1$ and $\mathbf{v}_2$) and eigenvalues (here $\lambda_1$ and $\lambda_2$), describing the independent directions and magnitudes of the variation of the data (see Fig. A.4B for a geometric illustration). Here, direction $v_1$ represents the variation in amplitude of the dose-response curves. For real data, typically most of the variation of the data is described by the largest three or four principal components. The small number of principal components can lead to a significant dimensionality reduction and data simplification.

## Chapter summary

- Variational principles are important tools for solving optimization problems. The Euler–Lagrange equation $\frac{\partial L}{\partial y} - \frac{\partial}{\partial x}\left(\frac{\partial L}{\partial y'}\right) = 0$ with $L = L(x, y, y')$ provides a general way of deriving equations of motions in physical systems.

- A mathematical model can be fitted to the data using maximum likelihood or Bayesian inference. Maximum likelihood maximizes the likelihood function, i.e. the conditional probability of the data given the model with certain parameter values. This boils down to minimizing the $\chi^2$ goodness of fit function of the sum of squared errors, weighted by the variance of the data. However, this can lead to overfitting. Instead, Bayesian inference maximizes the posterior probability, the conditional probability of obtaining the parameters given the data. Added benefits are the straightforward inclusion of prior knowledge and the return of uncertainty intervals on estimated parameters.

- Other mathematical tools include Fourier transform and principal component analysis. A Fourier transformation converts time-dependent behavior to frequency space, which can also help solve dynamical-system problems. Principal component analysis aims to find the linearly independent directions of fluctuations in data by diagonalizing the covariance matrix of the data. This shows the directions and magnitude of the variation in the data.

## Further reading

(1) *Statistical Thermodynamics of Nonequilibrium Processes*, by Joel Keizer (Springer Press, 1987).

(2) *Random Walks and Random Environments. Volume 1: Random Walks*, by Barry D. Hughes (Clarendon Press, Oxford, 1995).

(3) *Stochastic Processes in Physics and Chemistry*, by Nico G. Van Kampen (Elsevier, 1981).

(4) *Pattern Recognition and Machine Learning*, by Christopher M. Bishop (Springer, 2006).

(5) *Measurements and their Uncertainties: A Practical Guide to Modern Error Analysis*, by Ifan G. Hughes and Thomas P.A. Hase (Oxford, 2010).

(6) *Numerical Recipes*, by William H. Press, Saul A. Teukolsky, William T. Vetterling, and Brian P. Flannery (Cambridge University Press, 3rd Edition, 2007).

(7) *Statistical Mechanics*, by R. K. Pathria and Paul D. Beale (Elsevier, 3rd Edition, 2011).

(8) 'Variable sizes of Escherichia coli chemoreceptor signaling teams', by Robert G. Endres, Olga Oleksiuk, Clinton H. Hansen, Yigal Meir, Victor Sourjik, and Ned S. Wingreen, *Mol. Syst. Biol.* 4: 211 (2008).

# Index